Kinetic Modeling for Environmental Systems

Edited by Rehab O. Abdel Rahman

Published in London, United Kingdom

IntechOpen

Supporting open minds since 2005

Kinetic Modeling for Environmental Systems
http://dx.doi.org/10.5772/intechopen.79240
Edited by Rehab O. Abdel Rahman

Contributors
Ping Liu, Chunying Wang, Kanokporn Swangjang, Dina Hamad, Lizethly Caceres Jensen, Angelo Neira
Albornoz, Mauricio Escudey, Rehab O. Abdel Rahman

Notice
Statements and opinions expressed in the chapters are these of the individual contributors and not
necessarily those of the editors or publisher. No responsibility is accepted for the accuracy of
information contained in the published chapters. The publisher assumes no responsibility for any
damage or injury to persons or property arising out of the use of any materials, instructions, methods
or ideas contained in the book.

First published in London, United Kingdom, 2019 by IntechOpen
IntechOpen is the global imprint of INTECHOPEN LIMITED, registered in England and Wales,
registration number: 11086078, The Shard, 25th floor, 32 London Bridge Street
London, SE19SG – United Kingdom
Printed in Croatia

British Library Cataloguing-in-Publication Data
A catalogue record for this book is available from the British Library

Additional hard copies can be obtained from orders@intechopen.com

Kinetic Modeling for Environmental Systems
Edited by Rehab O. Abdel Rahman
p. cm.
Print ISBN 978-1-78984-726-0
Online ISBN 978-1-78984-727-7

We are IntechOpen,
the world's leading publisher of
Open Access books
Built by scientists, for scientists

4,100+
Open access books available

116,000+
International authors and editors

120M+
Downloads

Our authors are among the

151
Countries delivered to

Top 1%
most cited scientists

12.2%
Contributors from top 500 universities

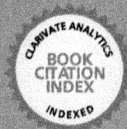

Interested in publishing with us?
Contact book.department@intechopen.com

Numbers displayed above are based on latest data collected.
For more information visit www.intechopen.com

Meet the editor

Rehab O. Abdel Rahman, Associate Professor of Chemical Nuclear Engineering at the Atomic Energy Authority of Egypt, was awarded a Nuclear Engineering PhD from Alexandria University. She has contributed to the publication of more than 30 peer-reviewed scientific papers, 11 book chapters, and seven books. She is an honored scientist of ASRT, serves as a verified reviewer in several journals, and is the managing editor for IJEWM and IJEE.

Contents

Preface

Interest in evaluating the impact of human activities on the environment has increased within the last decade. This interest is reflected in issuing stricter regulations and resources allocations to prevent and control potential negative impacts associated with different human activities. Environmental pollution is one of the adverse impacts of human activities; it is associated with historical and improper routine and accidental release of pollutants into the environment. In general, kinetic models are used to evaluate the driving forces that initiate temporal changes in any system and quantify these changes. These models are widely applied to design and optimize systems that support pollution prevention and control measures, i.e. different waste management activities and remediation projects. This book aims to present advances in developing and applying different kinetic models to support pollution prevention and control efforts. The authors summarize their experiences and present advances in different fields related to the presented topics. The book targets professional people in environmental industry and readers with technical backgrounds such as graduate and postgraduate students undertaking courses in environmental chemistry, ecology, and environmental engineering.

The book consists of three sections that cover important research and development efforts in modeling environmental systems. The first section introduces the assessment models as tools to support pollution prevention and control decisions. The editor describes the integration between the research and assessment models with special emphases on pollutant migration, presents the iterative nature of the assessment models, and explains the development of conceptual models by illustrating models that could be used to predict pollutant migration in different environmental subsystems. The chapter presents computational model selection and highlights simple models that could be used to estimate migration in terrestrial subsystems.

The second section highlights the development of kinetic models that could be used to support research efforts in preventing and controlling pollution generation. Prof. Liu and Dr. Wang present the development of a model that could be used to understand and analyze the physical mechanism and non-equilibrium condensation growth kinetics of carbon particles released from diesel engines. The chapter explains the condensation growth process and its control for soot particles using the Monte Carlo method. Dr. Hamad et al. develop a kinetic model to describe polyvinyl alcohol degradation in the advanced oxidation process. The model considers photolysis, polymer chain scission, and mineralization reactions to describe degradation, and the effect of the operating conditions are evaluated. The statistical moment approach was applied to model the molar population balance of live and dead polymer chains taking into account the probabilistic chain scissions of the polymer.

The third section displays environmental assessment studies for herbicide application, and development of a conceptual model for strategic environmental assessment. Prof. Jensen et al. present the results of a research effort to identify the features of ionizable and non-ionizable herbicides on volcanic ash-derived soils. The chemical and physical properties of both variable- and constant-charge soils are introduced and the sorption of metsulfuron-methyl onto both soil types is illustrated. Several models are tested to describe the sorptive behavior of ionizable and non-ionizable herbicides. Dr. Swangjang illustrates the development of a conceptual model to support strategic environmental assessment for mega projects. Consideration of the selection of the objectives, targets, and indicators is presented. A case study is presented considering the kinetic development resulting from changes in land use and ecological impacts are investigated. Finally, the conceptual model is presented.

I would like to thank cordially all the authors for their efforts that led to the production of this distinguished scientific contribution. An especial acknowledgment is directed to the author service manager, Ms. Marina Dusevic, for her coordination efforts.

Rehab O. Abdel Rahman
Atomic Energy Authority of Egypt,
Cairo, Egypt

Section 1

Introduction

Chapter 1

Introductory Chapter: Development of Assessment Models to Support Pollution Preventive and Control Decisions

Rehab O. Abdel Rahman

1. Introduction

The continuous increase in human activities affects the environment in notable ways; these effects need to be monitored and controlled when appropriate to ensure the sustainability of our lives. Environmental pollution is one of the major problems that associate these activities; it is initiated when a substance is released into the environment in a way that prevents its natural restoration [1, 2]. These releases could be classified as planned and uncontrolled releases. The first class is a part of routine human activity where discharge is performed after complying with the regulatory requirements, whereas uncontrolled releases associate accidents and nonregulated activities [1]. Uncontrolled releases and historical practices have led to several contamination problems, so restoration or remediation programs are being initiated to control these problems from spreading [2]. Currently, preventing and controlling environmental pollution and restoration of affected environmental systems receive great attention globally. This attention was translated into issuing strengthen regulations and allocating natural and human resources to support pollution prevention and control activities. In this respect, a continuous increase in research efforts was dedicated to investigate new materials and/or systems to evaluate their potential applications in preventing and controlling environmental pollution, that is, wastewater, gaseous, and solid waste management, and in and ex situ remediation projects. **Table 1** lists some pollution control and prevention systems and their classifications in terms of the scientific bases of the used technologies. These investigations are supported with enormous efforts to understand, simulate, predict, and decide on the performance of these materials and systems under predefined conditions using wide range of models. In this context, kinetic models are applied to:

1. assess the formation and/or evolution of the system and its subsystems;

2. assess, control, and optimize the chemical reactions used in different waste treatment technologies;

3. design and optimize the operation of remediation projects; and

4. support the decision-making process at regulatory agencies and operational facilities during different life cycle phases of pollution control and prevention systems, that is, planning, design, licensing, etc.

Technologies classification	Wastewater	Solid waste	Gaseous waste	Remediation In-& ex-situ
Physical	Sedimentation, Floatation.	Segregation, Compression, Shredding	Cyclone, Bag-House, Electrostatic precipitator	Soil washing, Soil vapor extraction
Physico-chemical	Solvent Extraction, Reverses osmosis Ultra & micro Filtration, Sorption/Ion Exchange, Coagulation/ Precipitation.	-	Stripping, Filters, Sorption	Permeable reactive barrier, Electro-Kinetic
Chemical	Advanced oxidation	-	-	Chemical Stabilization
Biological	Tricking filters, Attached growth on granular bio-filters, Activated sludge	Aerobic, Low/High- Anaerobic Digestion	-	Bio-treatment, Ex-situ-slurry biodegradation, Root zone Treatment
Thermal	Evaporation	Incineration	Combustion	Incineration, Vitrification

Table 1.
Technology for preventing and control of environmental pollution

Modeling by definition is an abstract of the real systems, where essential features, event, and process (FEP) that affect the performance of the studied system are presented [3, 4]. Generally, the modeling efforts are divided into research and assessment models. Research (process) models use laboratory and field experiments to identify FEPs that affect a subsystem or more, whereas assessment models link important processes (determined from research model) to predict the overall system performance [5, 6]. **Figure 1** illustrates the integration of research and assessment models, in which the studied subsystems are characterized and the factors that affect their behavior are identified experimentally. Then models are used within the research efforts to interpret, extrapolate/interpolate, and optimize the collected data; the modeling results will be used to evaluate and rank the FEPs that affect the system. In assessment models, important FEPs are linked to identify the problem formulation and basic system description, and then conceptual and computational models are constructed, verified, and used [5–11]. For instance, the quantification of the effect of time on the pollutants migration in terrestrial, aquatic, and/or atmospheric subsystems is usually conducted by measuring the concentration of major pollutants at incremental time at different distances from the source. Experiments are run for specified time determined based on the temporal scale of the study. The collected experimental data are analyzed to quantify the processes that control the migration. This analysis might include the use of simple empirical, semiempirical, or mechanistic mathematical models that allow a clear understanding of the nature of the processes that affect the migration. In terrestrial subsystems, these processes might include percolation, retardation, biodegradation, advection, and hydrodynamic dispersion [8, 11]. In subsequent sections, the development of assessment models to support the decision-making process will be illustrated with special emphasis on the prediction of pollutant migration. In this respect, the iterative nature of the assessment modeling will be overviewed, the conceptual model will be introduced, and some conceptual models that could be used to predict pollutant migration will be illustrated. The selection of

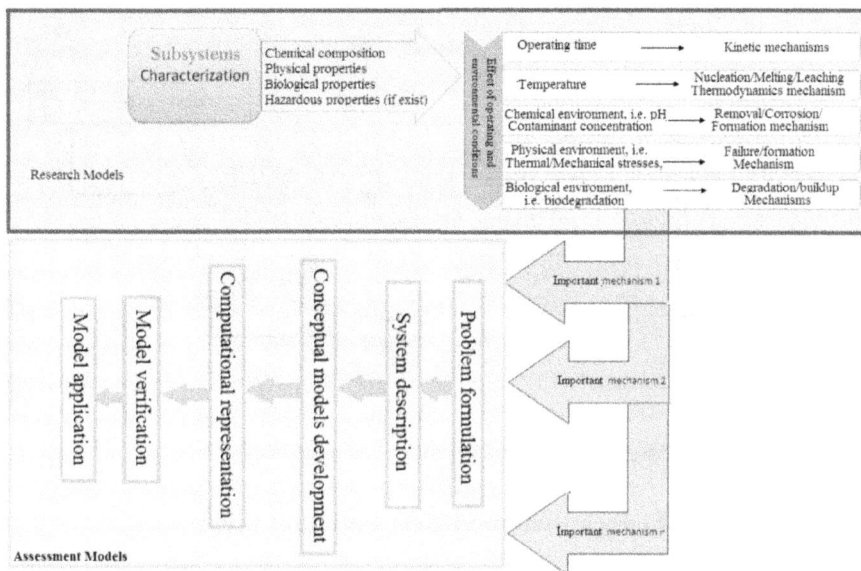

Figure 1.
Integration of research and assessment models in studying a system.

computational models will be presented, where some simple models that could be used to estimate the migration in terrestrial subsystems will be summarized.

2. Iterative nature of the assessment modeling

Assessment models are used to support the decision-making process during different life cycle stages of any pollution prevention and/or control system, for example, sitting waste management facility and designing remediation program. Their outputs should provide assurance that the systems will be sited, designed, operated, etc., in a manner that compiles with the safety requirement issued by the regulatory body. Assessment modeling starts with problem formulation and basic system description based on available system information. During problem formulation, the assessment objectives and audiences, regulatory framework, system boundaries, spatial and temporal scales, stage of project development, critical receptors (affected groups), adopted assessment approaches, nature of assumptions, data availability, level of accuracy, cost, and uncertainty treatment should be clarified [4]. The level of the assessment complexity is largely dependent on the national regulations and state of project development. Assessment modeling is an iterative process, where basic system data are used to develop a simple model that contains all essential FEPs derived based on basic system description. The model is then verified using system-specific data to check its prediction adequacy. If adequate simulation results are obtained, the model will be applied; otherwise more system-specific data should be collected to help in improving the model predictions. **Figure 2** illustrates the iterative nature of the modeling process and its relation with the system-specific data, in which the developed model complexity or simplicity is determined based on the stage of development of the studied system and the availability of system-specific data [11, 12]. The developed model, in each iterative stage, is produced from multi-step process that includes the development of conceptual and computational models (mathematical model and the tool that solves the mathematical model) [5–9].

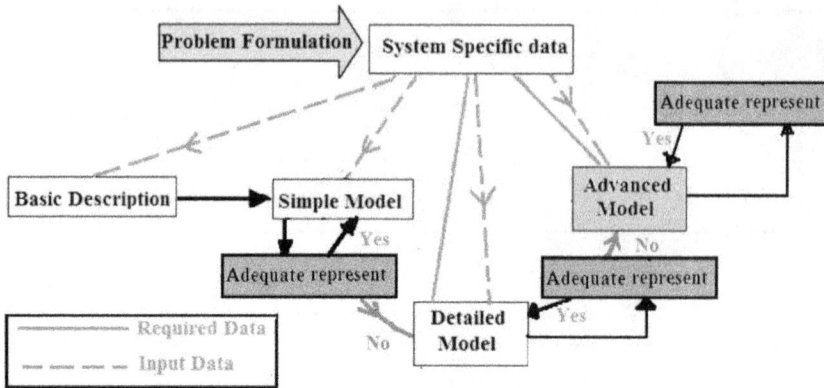

Figure 2.
Iterative nature of the modeling process and its relation with system-specific data.

3. Development of conceptual model for pollutant migration assessments

Conceptual model is defined as "A simplified representation of how the real system is believed to behave based on a qualitative analysis of field data" [11]. The development of a conceptual model starts with a clear determination of available information and knowledge gaps about the system. Subsequently, essential FEPs and their interactions in each subsystem are identified, and assumptions that were made to include or exclude any of these FEPs are highlighted based on the results of the research models [11]. Finally, flowcharts are used to describe the graphical relationship between different processes in different physical subsystems. It should be noted that the conceptual model could be imperfect if over- or under-simplification of the studied system were used, where over-simplification can lead to ineffective model with large uncertainties and under-simplification can lead to complex model that raises the project costs. Imperfect conceptual model could be resulted from incomplete problem identification/assessment context, wrong assumptions in developing the conceptual model, and poor identification of the important processes.

Conceptual models are usually constructed based on source-pathway-receptor analysis, where pollution sources are defined by investigating the driving forces and duration of the releases for each pollutant, the routes of pollutant transport between different physical subsystems are determined, and receptor exposure mechanisms and duration are identified [9, 13, 14]. Below are some examples that illustrate the construction of conceptual model for pollutant migration into different subsystems that could be developed to support the pollutant control and prevention decision-making process.

To characterize the extent of the contamination problems due to contaminant spill, there is a need to collect samples from potentially affected subsystems, that is, groundwater, surface water, air, and soil and subsoil. Sampling procedure should consider both the main pollutants and subsystem properties, for example, pollutant concentrations in different subsystems, water pH, velocity, wind velocity, etc. Characterization results will be analyzed within the research modeling efforts, and the results of this analysis will determine the complexity of the model. Based on these results, homogenous or nonhomogenous subsurface may be considered to estimate pollutant percolation and sorption, and the elimination or inclusion of biodegradation and aquifer recharge as sink or source for pollutants in the subsurface and surface water will be determined. In this case, different terrestrial and atmospheric exposure pathways to receptors, downstream the contamination source, were identified as main exposure routes. **Figure 3** illustrates the main processes that can lead to pollutant migration or attenuation from a contaminant spill into different subsystems. The pollutants are assumed to be transported by percolation, surface runoff, and evaporation, and attenuation is assumed to occur as

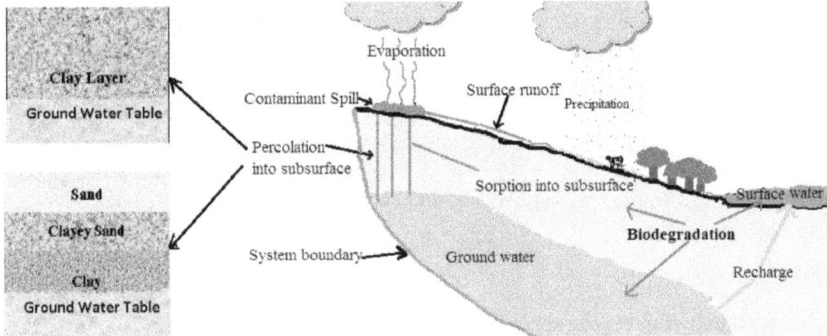

Figure 3.
Conceptual model to predict pollutant migration/attenuation from the source term into the surrounding environment.

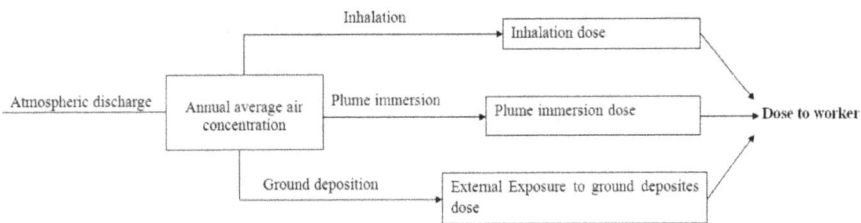

Figure 4.
Conceptual model to quantify the effect of continuous atmospheric discharge on the worker [14].

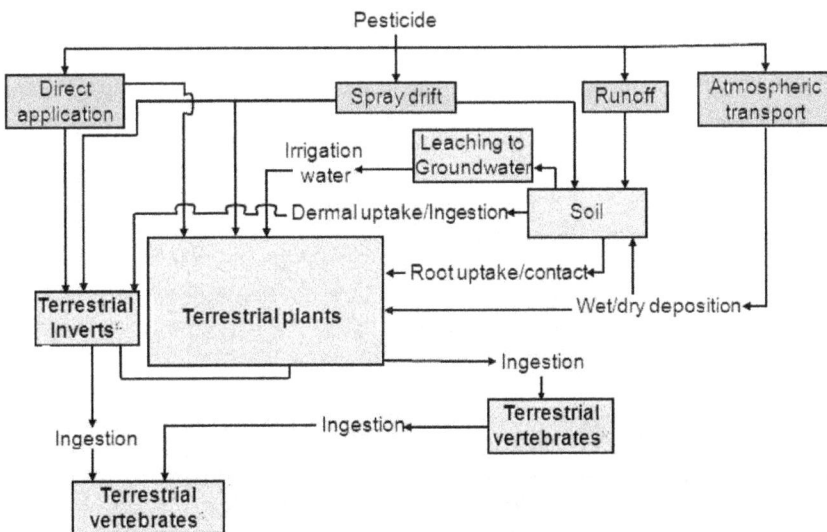

Figure 5.
Conceptual model to quantify the effect of pesticide application on the environment [15].

a result of sorption into the subsurface and biodegradation within surface water, groundwater, and geosphere.

To determine the worker dose in a radioactive waste incinerator facility during the planning phase for transition from batch to continuous operation, a conceptual model was constructed [14]. The pollutants are assumed to be transported through the air via advective-diffusive process, and the exposure means were determined to include inhalation of gaseous pollutants (which is the main exposure mean in that study), direct dermal exposure, and ingestion of contaminated water (**Figure 4**).

Generic conceptual model to quantify the effect of pesticide application on the environment is suggested by US EPA (**Figure 5**) [15]. The model represents terrestrial exposure pathways, where the pollutants (pesticide) are transported through the atmospheric and aquatic subsystems and were assumed to affect terrestrial receptors, that is, plants, invertebrates, and vertebrates. The exposure means included inhalation, dermal exposure, and ingestion with a detailed characterization of the dietary routes.

4. Computational representation of the conceptual model

The development of the computational model that represents accurately the conceptual model is a crucial task, where the accuracy of the obtained results will be used to judge if the modeling effort is enough to represent the system or there will be a need to acquire field data and develop an updated model (**Figure 2**). For a simple conceptual model, a simple empirical model could be used, as the site-specific information is available and a more realistic model could be used [13]. The type of the mathematical representation of the conceptual model is defined during the problem formulation, and the selection of the appropriate model is bounded by [4, 11]:

1. System dimensions: decision should be made if one, two, and three dimensions will be used to represent the system.

2. Nature of the boundary conditions: Source terms release assumptions should identify if the release is constant or variable throughout the time and space.

3. Steady state or time variant model: the system behavior is changing with time or fixed.

4. Uncertainty management: probabilistic or deterministic approaches.

5. Homogenous and nonhomogenous system.

6. Type of flow and transport process: the flow occurs via intergranular or fissure flow, and the transport is governed by advection or hydrodynamic dispersion.

During the development of a mathematical representation, the studied system is usually divided into a subsystem. For the conceptual model presented in **Figure 3**, the system could be divided into source subsystem which describes the mobilization of the pollutant from the source, terrestrial migration, atmospheric transport, and receptors subsystems. **Table 2** shows some simple models that could be used to develop a mathematical representation of pollutant migration in terrestrial compartment [5, 16–20]. This table presents models that could be used to estimate both

Model use	Parameters	Model
Infiltration rate in homogenous soil, (q, m/d)	Soil sorptivity (S, m/d$^{0.5}$), Soil dependent constant (A)	$q(t) = \frac{1}{2}St^{-0.5} + A$
Flow rate in homogenous soil, (q, m/d)	Hydraulic gradient (i), Hydraulic conductivity (k, m/d)	$q = ki$
Flow rate in non-homogenous soil (q, m/d)	Dimensionless time (t*), Dimensionless depth (z*), Change in volumetric water content as the wetting front passes layer n ($\delta\theta$, m^3/m^3), Potential head while the wetting front passes through layer n, (H$_n$, m)	$q = \left(\dfrac{0.5\left[t^* - 2z^* + \sqrt{(t^* - 2z^*)^2 + 8t^*}\right] + 1}{0.5\left[t^* - 2z^* + \sqrt{(t^* - 2z^*)^2 + 8t^*}\right] + z}\right) * k_n$ $t^* = \dfrac{k_n t}{\delta\theta\left(H_n + \sum_{i=1}^{n-1} Z_i\right)}$ $z^* = \dfrac{k_n}{\left(H_n + \sum_{i=1}^{n-1} Z_i\right)} \sum_{i=1}^{n-1} \dfrac{z_i}{k_i}$
Pollutant Travel time, (t, d)	Vadose zone thickness (d, m), Porosity (n).	$t = dn/q$
Water average velocity (v, m/d)		$v = Ki/n$
Hydrodynamic dispersion (Dl, m^2/d)	Effluent pore volume (u), Distance (L, m), Mean pore water velocity (v, m/d).	$D_i = \frac{vL}{8}\left(\frac{U-1}{\sqrt{U}}\Big|_{0.84} - \frac{U-1}{\sqrt{U}}\Big|_{0.16}\right)^2$
Distribution coefficient (kd) of element (i) assuming linear isotherm	Concentration in the solution (C, ppm) at initial (i) and final (e) state, Solution volume (V, l), Soil weight (m, g)	$K_{di} = \left(\frac{C_{ii} - C_{ei}}{C_{ii}}\right)\left(\frac{V}{m}\right) \times 1000$
Retardation coefficient (Rf) in vadose zone	Soil density (ρ, kg/m^3), Soil porosity (ε).	$R_f = 1 + \frac{\rho(1-\varepsilon)}{\theta} K_{di}$
Retardation factor assuming Freundlich isotherm	Freundlich constant indicative of the relative sorption capacity (n) and (Kf, mg/g)	$R_f = 1 + \frac{\rho K_f}{\theta n} C^{\left(\frac{1-n}{n}\right)}$
Retardation factor assuming D-R	Maximum sorbed as calculated by D-R isotherm (qm, mg/g), Energy of sorption estimated by D–R model (E, kJ/mol), Gas constant (R,8.314 J/mol K), Absolute temperature (T, K)	$R_f = 1 + \frac{\rho RT q_m E^2}{\theta} \exp\left(\frac{RTln(1+1/c)^2}{2E^2}\right)$ $ln\left(\frac{C+1}{C}\right)\left(\frac{C}{C+1}\right)\left(\frac{1}{C}\right)$

Table 2.
Mathematical models used to assess the migration in soil subsurface [5, 16–20].

flow (infiltration/flow rate, travel time, and average water velocity) and transport parameters (hydrodynamic dispersion, distribution, and retardation coefficient) for homogenous and nonhomogenous soil under saturated and vadose conditions.

Acknowledgements

The author would like to acknowledge Dr. A.A. Zaki, professor of nuclear chemical engineering at Atomic Energy Authority of Egypt, for the time and efforts that he spent to review this work.

Author details

Rehab O. Abdel Rahman
Hot Lab. Center, Atomic Energy Authority of Egypt, Cairo, Egypt

*Address all correspondence to: alaarehab@yahoo.com

IntechOpen

References

[1] Abdel Rahman RO, Kozak MW, Hung Y-T. Radioactive pollution and control, Ch (16). In: Hung YT, Wang LK, Shammas NK, editors. Handbook of Environment and Waste Management. Singapore: World Scientific Publishing Co; Feb 2014. pp. 949-1027. DOI: 10.1142/9789814449175_0016. Available from: http://www.worldscientific.com/doi/abs/10.1142/9789814449175_0016

[2] Abdel Rahman RO, Elmesawy M, Ashour I, Hung Y-T. Remediation of NORM and TENORM contaminated sites–review article. Environmental Progress & Sustainable Energy. 2014; 33(2):588-596

[3] Chen Q, Han R, Ye F, Li W. Spatio-temporal ecological models. Ecological Informatics. 2011;6:37-43

[4] IAEA. Safety Assessment Methodologies for Near Surface Disposal Facilities, vol. 1. Vienna: IAEA; 2004. ISBN: 92-0-104004-0

[5] Abdel Rahman RO, El-Kamash AM, Zaki AA. Modeling the long term leaching behavior of 137Cs, 60Co, and152,154Eu radionuclides from cement-clay matrices. Hazardous Materials. 2007;145:372-380

[6] Drace Z, Mele I, Ojovan MI, Abdel Rahman RO. An overview of research activities on cementitious materials for radioactive waste management. Materials Research Society Symposium Proceedings. 2012;1475:253-264. DOI: 10.1557/opl.2012

[7] Abdel Rahman RO, Rakhimov RZ, Rakhimova NR, Ojovan MI. Cementitious Materials for Nuclear Waste Immobilisation. New York: Wiley; 2014. ISBN: 9781118512005. http://eu.wiley.com/WileyCDA/WileyTitle/productCd-1118512006,subjectCd-CH50.html

[8] Abdel Rahman RO, Ibrahim HA, Abdel Monem NM. Long-term performance of zeolite Na A-X blend as backfill material in near surface disposal vault. Chemical Engineering Journal. 2009;149:143-152

[9] Abdel Rahman RO, Saleh HM. Introductory chapter: Safety aspects in nuclear engineering. In: Abdel Rahman RO, Saleh HM, editors. Principles and Applications in Nuclear Engineering: Radiation Effects, Thermal Hydraulics, Radionuclide Migration in the Environment. IntechOpen: DOI: 10.5772/intechopen.76818. ISBN: 978-1-78923-616-3. Available from: https://www.intechopen.com/books/principles-and-applications-in-nuclear-engineering-radiation-effects-thermal-hydraulics-radionuclide-migration-in-the-environment/introductory-chapter-safety-aspects-in-nuclear-engineering

[10] Guidance on risk assessment and the use of conceptual models for groundwater. Guidance Document No. 26. ISBN-13: 978-92-79-16699-0. 2010. DOI: 10.2779/53333

[11] McMahon A, Heathcote J, Carey M, Erskine A. Guide to Good Practice for the Development of Conceptual Models and the Selection and Application of Mathematical, National Groundwater & Contaminated Land Centre Report NC/99/38/2 (2001). ISBN: 1 857 05610 8

[12] Abdel Rahman RO. Preliminary evaluation of the technical feasibility of using different soils in waste disposal cover system. Environmental Progress & Sustainable Energy. 2011;30(1):19-28

[13] Barnthouse LW, Suter GW II, Guide for Developing Data Quality Objectives for Ecological Risk Assessment at DOE Oak Ridge, Operations Facilities, ES/ER/TM-185/R1; Springfield, VA: National

Technical Information Service, U.S. Department of Commerce; 1996

[14] Abdel Rahman RO. Preliminary assessment of continuous atmospheric discharge from the low active waste incinerator. International Journal of Environmental Sciences. 2010;**1**(2): 111-122

[15] EPA. Guidance for the Development of Conceptual Models for a Problem Formulation Developed for Registration Review. Available from: https://www.e pa.gov/pesticide-science-and-asse ssing-pesticide-risks/guidance-deve lopment-conceptual-models-problem. [Accessed: 28/11/2018]

[16] Abdel Rahman RO, Abdel Moamen OA, Hanafy M, Abdel Monem NM. Preliminary investigation of zinc transport through zeolite-X barrier: Linear isotherm assumption. Chemical Engineering Journal. 2012;**185–186**: 61-70

[17] Abdel Rahman RO. Performance assessment of unsaturated zone as a part of waste disposal site [PhD thesis]. Egypt: Nuclear Engineering Dep., Faculty of Engineering, Alexandria University; 2005

[18] Abdel Rahman RO, El Kamash AM, Zaki AA, El Sourougy MR. Disposal: A last step towards an integrated waste management system in Egypt. In: International Conference on the Safety of Radioactive Waste Disposal; Tokyo, Japan. IAEA-CN-135/81; 2005. pp. 317-324

[19] Gasser MS, El Sherif E, Abdel Rahman RO. Modification of Mg-Fe hydrotalcite using Cyanex 272 for lanthanides separation. Chemical Engineering Journal. 2017;**316C**:758-769

[20] Abdel Rahman RO, Ibrahim HA, Hanafy M, Abdel Monem NM. Assessment of synthetic zeolite NaA-X as sorbing barrier for strontium in a

radioactive disposal facility. Chemical Engineering Journal. 2010;**157**:100-112

Section 2

Pollution Prevention
and Controls

Chapter 2

The Diesel Soot Particles Fractal Growth Model and Its Agglomeration Control

Ping Liu and Chunying Wang

Abstract

Based on the fractal growth physical model of soot particles from large diesel agriculture machinery, this chapter simulates the morphological structure of collision for the single particles and single particles, single particle and clusters, clusters and clusters, firstly. Moreover, combining with the collision frequency, the fractal growth is controlled to agglomeration using the main environmental factors interference for diesel engine soot particles, in order to make them condensed into regular geometry or larger density particles, reduce the viscous drag for capturing by the capturer or settlement and to realize the control of the pollution of the environment. The results of numerical simulation show that the proposed method is feasible and effective, which will help to understand and analyze the physical mechanism and kinetics of non-equilibrium condensation growth behavior of the actual carbon smoke particles and provide the solution to further reduce emissions of the inhalable particulate matter from diesel engines.

Keywords: soot particles, agglomeration, fractal growth, control, diesel engine

1. Introduction

Large diesel agriculture machinery plays an important role in economic development, but it also brings sharp problems in environmental protection. Diesel emissions are one of the most important sources of air pollution. There are many kinds of harmful substances, such as HC, CO, NOX, and soot particles, but the emission of harmful gases from diesel engines, such as HC and CO, is quite low; NOX emissions are also in the same order of magnitude as gasoline engines. The soot particles emitted are respirable particles, causing the most serious air pollution [1, 2].

Particles emitted by diesel engines are usually composed of soot, organic soluble components, and sulfides [3]. The main components of particles discharged from a typical heavy-duty diesel engine under transient conditions are shown in **Figure 1**. Usually, soot accounts for 50–80% of total particulate matter. It is one of the most important harmful emissions [3]. Therefore, it is of great significance to control the emission of soot particles from diesel engine emissions.

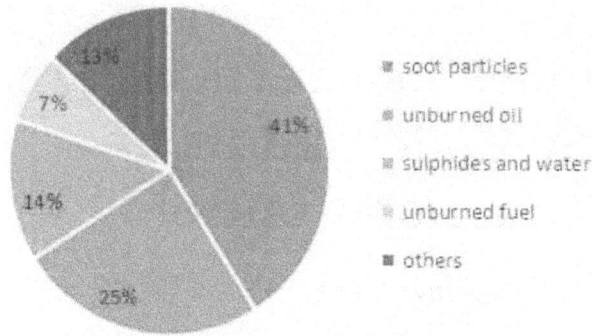

Figure 1.
The composition of particulate emissions from the heavy-duty diesel [3].

Soot is a very fine particle formed by a complex reaction mechanism in the flame of the fuel-rich region when burning hydrocarbons in the absence of air, mainly composed of a mixture of amorphous carbon and organic matter [4]. Since the concentration and particle size of soot particles emitted by the gasoline engine is lower than that of the diesel engine [5], this chapter mainly analyzes the soot particles of the diesel engine.

At present, the study of soot particles in diesel engines has focused on optical properties, chemical composition, particle size distribution, source analysis, and human health assessment [6, 7], but research on particle morphology (morphology and surface structure) is almost blank, especially the morphological structure of the particles. Most soot particles have complex fractal morphology [8, 9], affecting the nature of the particles. By studying its fractal structure, the deposition of particles, the viscous resistance of the particles and the adsorption of toxic molecules can be deduced. Therefore, it is necessary to control the fractal condensation and growth morphology of diesel soot particles.

Based on the fractal growth physical model of soot particles from large diesel agriculture machinery, this chapter simulates the morphological structure of collision for the single-single particles, single-clusters, clusters-clusters, firstly. Moreover, combining with the collision frequency, the fractal growth is controlled to agglomeration using the main environmental factors interference for diesel engine soot particles, in order to make them condensed into regular geometry or larger density particles, reduce the viscous drag for capturing by the capturer or settlement and to realize the control of the pollution of the environment.

2. The condensation growth process of the soot and its simulation method

2.1 The generation process of soot

The soot particles generation process undergoes complex chemical reactions and physical processes. Firstly, it undergoes gas phase reactions, phase transitions from gaseous to solid state. Then the formation of soot particles in diesel cylinders undergoes the evolution of kinetic events such as nucleation, condensation, collision fragmentation, growth, and surface oxidation [10–14]. The specific formation process described by the soot particle model is shown in **Figure 2**.

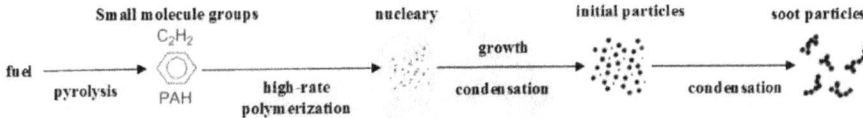

Figure 2.
The soot generation process.

2.2 The simulation method of soot structure

According to the characteristics of the soot growth process, the dynamic Monte Carlo method [15] is used to establish the soot fractal growth model. As shown in **Figure 3**, in a two-dimensional Euclidean space with many particles, one initial particle is set as the target particle, and the other particles are candidate particles. One of the candidate particles is selected to collide with the target particle according to a randomly generated locus, and adheres according to the adhesion probability. One other candidate particle repeats the above process, and the analogy eventually forms an agglomerate. If the motion reaches the boundary of the space, the particle is absorbed by the target particle and disappears. After the particles are released, they do Brownian motion, and they are required to move to the neighboring left, right, upper, and lower surrounding squares with a probability of 1/4. The process will continue until the particles leave the boundary or reach the agglomerate. There are two kinds of collision for particles: the collision of the particle with particle (**Figure 4**), the collision of the cluster with cluster (**Figure 5**).

Taking the collision of the single particle as an example to illustrate the collision of the particles, the trajectory vectors of two collision particles is firstly determined

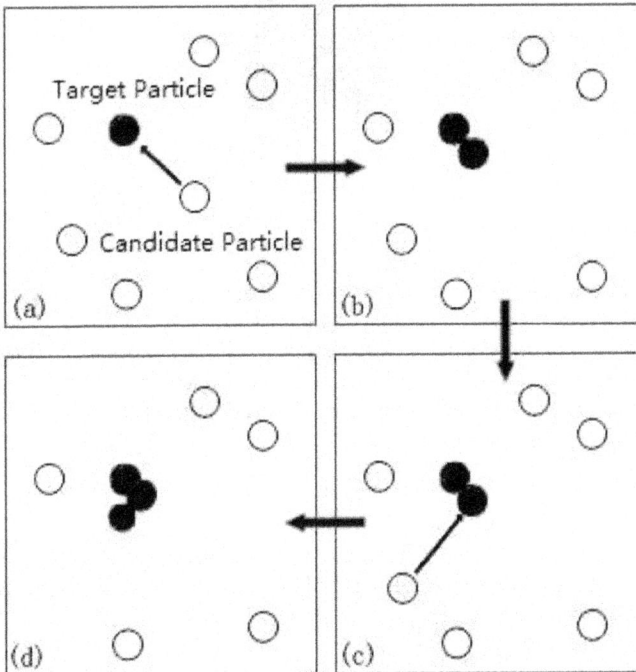

Figure 3.
The diagram of growth process. ((a) set target particle; (b) collide with target particle; (c) analogy of collide with target particle; (d) form an agglomerate).

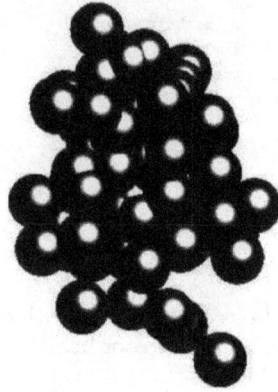

Figure 4.
The collision of single particles and single particle.

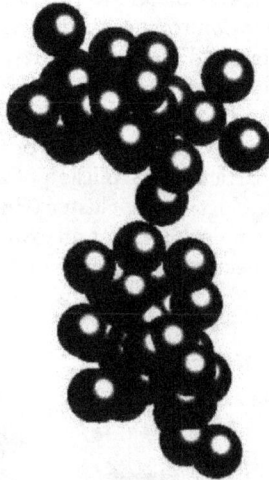

Figure 5.
The collision of clusters and clusters.

in **Figure 6**. Two small balls are defined as B_1 (target particles) and B_m (random particles) to represent two separate particles, with radius R_1 and R_m, respectively. The coordinates of B_1 are given, the coordinates of B_m are random, and the radius of the concentric sphere B_s of the small ball B_1 is defined as R_s ($R_s = R_1 + R_m$). Then let B_m move according to a random trajectory. When B_m meets the fixed ball B_1, Eq.(1) is satisfied, where x_s^0 is the center of the ball B_1.

$$|x_s - x_s^0| = R_s. \tag{1}$$

When the random particle B_m adsorbs and condenses on the target particle B_1, its random motion trajectory vector v exactly intersects with or intersects the ball B_1, and can be defined as Eq.(2),

$$x_m = x_m^0 + c_n v, \qquad c_n \in [0, \infty). \tag{2}$$

where x_m is the sphere center coordinate after the collision of the ball B_m, and x_m^0 is the initial sphere center coordinate of the ball B_m.

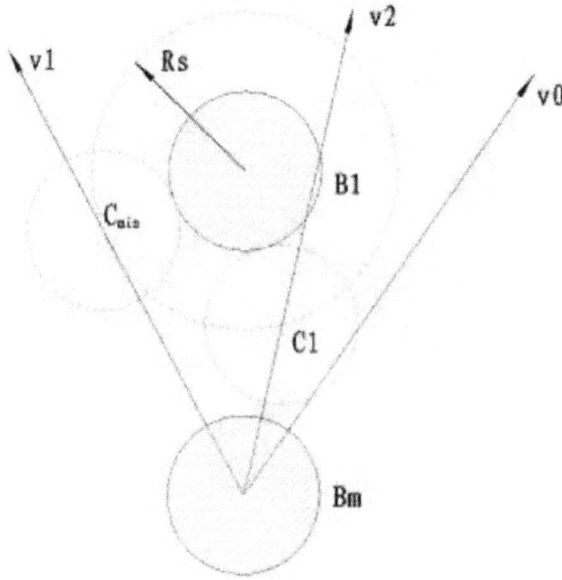

Figure 6.
The diagram of single particle collisions.

The collision of two small balls can be defined Eq.(3).

$$|x_m - x_m^0| = |(x_m^0 - x_s^0) + c_n v| = R_s.$$

(3)

Solve both sides of squared Eq.(3) simultaneously to obtain a quadratic equation with unknown c_n. c_n is the judgment factor, if c_n has no solution, two balls cannot collide; if c_n has a solution, the two balls collide with two cases. One is that when there is a unique solution, the two balls just collide with each other; when there are two solutions, the smallest solution is c_{min} according to the physical conditions. Then the coordinates of randomly moving ball B_m are also determined as Eq.(4). This process describes a simple collision process between particles and particles. Based on this, it can be used to simulate the collision and clustering process between clusters and clusters. The collision process is still established by using Monte Carlo method.

$$x_m = \begin{cases} x_m^0 + c_{min} v, & \text{intersect,} \\ x_m^0 + c_{min} v, & \text{tangent.} \end{cases}$$

(4)

The shape of aggregates formed by fractal growth of soot is closely related to the radius of gyration, soot radius, and fractal dimension, which is shown in **Figure 7**. The R_g characterizes the compactness of soot condensation growth, R_e characterizes the size of the soot agglomerates formed by fractal growth. The fractal dimension of the surface roughness of soot agglomerates is closely related to the adsorption of particles. The calculation of fractal dimension D_f of soot agglomerates is based on the box calculation method [16] and shown in Eq.(5), where $N_n(A)$ is the minimum number of boxes needed to contain A, $1/T_n$ is the boundary of the small box. When T_n is large enough, the box dimension is approximate as Eq.(6), and the fractal dimension calculation method for soot agglomerates is shown in **Figure 7**.

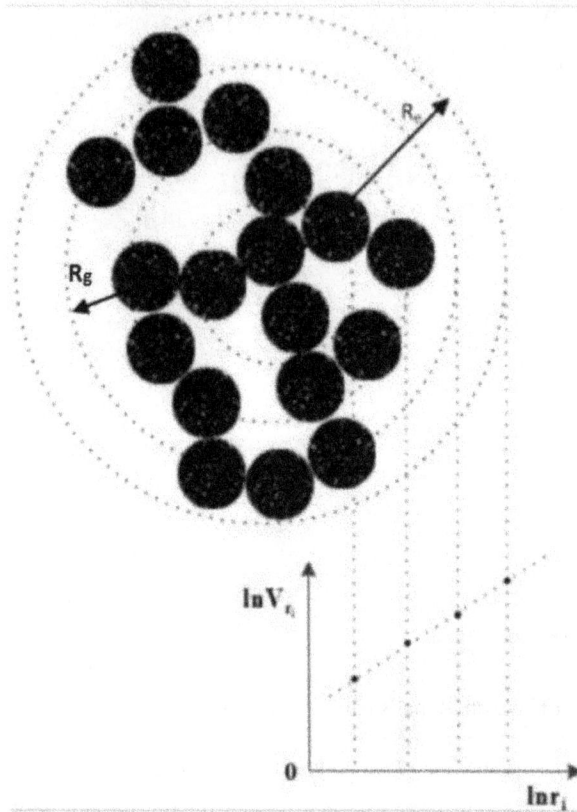

Figure 7.
The sandbox method for the fractal dimension of soot condensed matter.

$$D_f = \lim_{n \to \infty} \frac{N_n(A)}{-ln T_n}, \tag{5}$$

$$D_f = slope\{\ln(N_n(A)), \ln(T_n)\}. \tag{6}$$

3. The analysis and condensation control of soot particles fractal growth

3.1 The theory of control

The formation of soot particles in diesel engines is affected by factors such as temperature, pressure, soot particle concentration, and oxidation rate [17]. According to the characteristics of free particles moving in a continuous medium state, it can be considered that the soot growth of a soot particle (condensation collision) has a distribution parameter equation of motion with boundary conditions as Eq.(7).

$$\frac{\partial \eta(x,y,t)}{\partial t} = \delta\left(\frac{\partial \eta^2(x,y,t)}{\partial x^2} + \frac{\partial \eta^2(x,y,t)}{\partial y^2}\right), \tag{7}$$

The condensation growth frequency of soot particles satisfies the distribution parameter system Eq.(8).

$$\nabla^2 \eta(x,y) = F\left(\eta(x,y), \frac{\partial \eta}{\partial t}, u(x,y)\right), \tag{8}$$

$\eta(x,y)$ is the condensation temperature. F represents the environmental distur-bance term, called the forcing term, which is a non-linear function term. $u(x,y)$ is the initial value of the digitization, called the source term. The solution to the system Eq.(8) is very tedious. To facilitate the analysis of the solution, a discrete power system of Eq.(8) is introduced as Eq.(9).

$$\eta_{m+1,n} + \eta_{m-1,n} + \eta_{m,n+1} + \eta_{m,n-1} - 4\eta_{m,n} = F\big[\eta_{m,n}, (\eta_{m+1,n} - \eta_{m,n})(m_{t+1} - m_t)$$
$$+ (\eta_{m,n+1} - \eta_{m,n})(n_{t+1} - n_t), u_{m,n}\big], \tag{9}$$

Considering the boundedness and variability of soot particle agglomeration, the nonlinear function F is set as Eq.(10).

$$F = \alpha \sin(\eta_{m,n}) + u_{m,n}, \tag{10}$$

For more generalized processing problems, the system Eq.(11) is hereby introduced.

$$\Omega(r) = \alpha \sin(\Omega(r-1)) + \Omega(r-1) + u_{m,n}, \tag{11}$$

where

$$\Omega(r) = \eta_{m+r,n} + \eta_{m-r,n} + \eta_{m,n+r} + \eta_{m,n-r}, r = 1, 2, \cdots. \tag{12}$$

Obviously, when $r = 1$, Eq.(12) becomes system Eq.(9). By iterating simplifica-tion of Eq.(9), a simple control system can be obtained as Eq.(13).

$$\Omega(r) = ru + \alpha \sin(\Omega(r-1)) + \alpha \sin(\Omega(r-2)) + \cdots + \alpha \sin(\Omega(0)) + \Omega(0), r = 1, 2, \cdots \tag{13}$$

3.2 The control of fractal growth for diesel engines' soot particles from source item and nonlinear term

According to the control method of [18], this chapter analyzes the effect of this control method on the fractal growth of soot particles. Assuming that $\mathcal{H}$ is a condensed region, $\overline{\mathcal{H}}$ is a condensed boundary, and $\mathcal{M}$ is the scope of control of the source item $u(x,y)$, and satisfies $Mathcal\,M \in \mathcal{H}$. In addition, for any $(x,y) \in \mathcal{H}$, there is $0 \leq \eta(x,y) \leq 1$ established. Since the analytical function $u(x,y)$ satisfies the maximum principle in $\mathcal{H}$, for any $(x,y) \in \mathcal{H} - \overline{\mathcal{H}}$, condition $0 \leq \eta(xy) < 1$ must be true, so α and u in Eq.(11) must be as small as possible, represented by an inequality:

$$0 \leq \eta_{m+r,n} + \eta_{m-r,n} + \eta_{m,n+r} + \eta_{m,n-r} < 1,$$

$0 \leq \Omega(r) < 1$, where $r = 1, 2, \cdots$. Since $0 \leq \sin(\Omega(t)) < \Omega(r) < 1$ holds, the system (10) satisfies the relationship Eq.(14).

$$\Omega(r) < ru + \alpha\Omega(r-1) + \alpha\Omega(r-2) + \cdots + (\alpha+1)\Omega(0). \tag{14}$$

According to Eq.(13) and combined with mathematical induction, Eq.(15) can be get.

$$\Omega(r) < \left[(\alpha+1)^{r-1} + (\alpha+1)^{r-2} + \cdots + (\alpha+1) + 1\right]u + (\alpha+1)^r\Omega(0), r = 1, 2, \cdots.$$

$$(15)$$

Eq.(16) is get by changing the inequality of $\Omega(r)$ to $\Psi(\alpha, u, r)$

$$\Psi(\alpha, u, r) = \left[(\alpha+1)^{r-1} + (\alpha+1)^{r-2} + \cdots + (\alpha+1) + 1\right]u + (\alpha+1)^r\Omega(0) \quad (16)$$

And because $0 < \alpha \leq 1$, $0 < u \leq 1$, it turns out to have $\frac{\partial\Psi}{\partial\alpha} > 0$ and $\frac{\partial\Psi}{\partial u} > 0$ is true, $\Omega(r)$ is monotonically increasing about α and u, respectively.

The particle condensation temperature η will increase with the increase of the nonlinear term $\alpha \sin(\eta)$ and the source term u. For system Eq. (8), when the action region of source item u is circular (r is a radius), the values of u are constants and random numbers(rand represents a random number in the range (0,1), respectively, and the resulting simulated pictures are shown in **Figures 8–10**. Comparing with **Figure 3** and **Figure 4** without interference and other model [19] shown in **Figure 11**, this chapter simulates the morphological structure of collision for the single-single particles, single-clusters, clusters-clusters, and it is obvious that the effect of the increase of the interference term and the action region on the control of the aggregation of particles is more and more condensed than in the absence of the

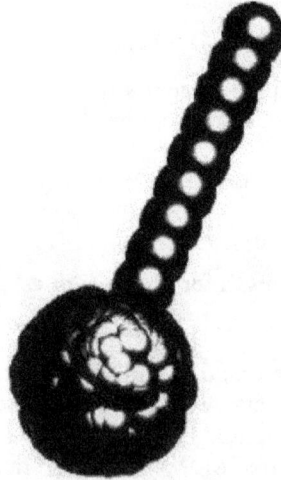

Figure 8.
The control of single direction.

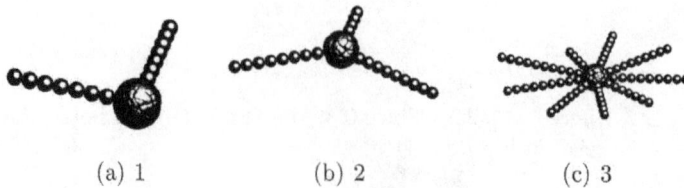

(a) 1   (b) 2   (c) 3

Figure 9.
The control of multiple direction.

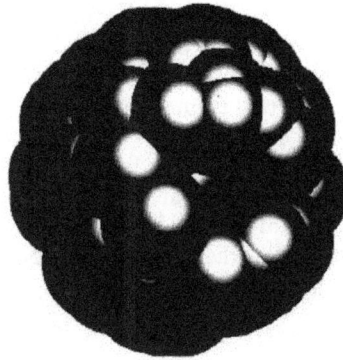

Figure 10.
Particles' center point coagulation control.

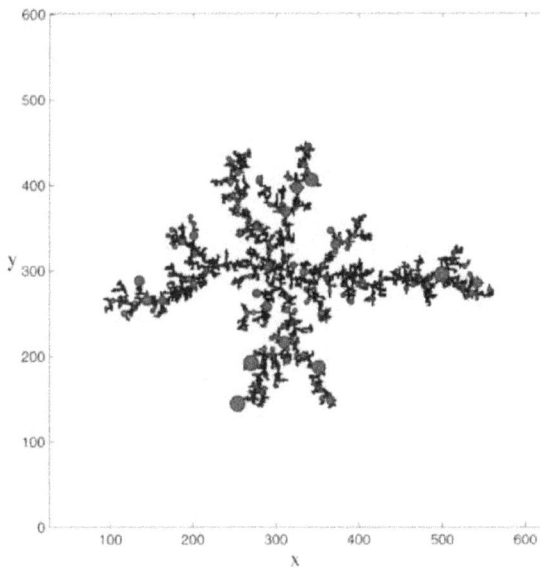

Figure 11.
Fractal diffusion of soot particles established by others [19].

interference term. The concentration of the particles after the condensation is greater for the settlement and the aggregation. Condensation can also have a fixed direction. The control method on the fractal growth reduces the complexity of the surface area of aggregated particles and reflects the effectiveness of the control method.

4. The meaning of the soot particles condensed control

The fractal structure of the particles is closely related to the binding resistance and the adsorption of the particles. The literature [20] investigated the relationship between the viscous resistance and the fractal structure of the particles during the descending process. Particles with a fractal structure will have a larger fractal dimension, the smaller the viscous resistance, and the faster the sedimentation rate than spherical particles of the same volume. The fractal dimension of the particle before control shown in **Figures 3** and **4** is 2.029 and 2.236, respectively. The fractal dimension of the particle after control in **Figure 8** is 2.3273. Obviously, the viscous resistance of the particle in **Figure 8** is small, which is conducive to the settlement of particles.

The fractal dimension of particles directly affects the surface adsorption. The relationship between the number of saturated molecules adsorbed in single layer N_m and the cross-sectional area of adsorbed molecules S_m is given by Eq. (17), where ξ is the scale factor and D_f is the fractal dimension of the particle.

$$N_m = \xi(S_m)^{\left(-\frac{D_f}{2}\right)},\tag{17}$$

If the adsorbate molecular weight is M and the density is ρ, then the adsorption amount Q' is Eq.(18).

$$Q' = N_m \frac{M}{\rho},$$
$$Q' = \varepsilon \frac{M}{\rho}(S_m)^{\left(-\frac{D_f}{2}\right)}.\tag{18}$$

Obviously, the adsorption of toxic particulates by atmospheric particles is not only related to the composition and chemical properties of gas molecules but also related to the fractal dimension of the particle surface. The roughness of the surface of atmospheric particles also affects the adsorption of toxic gases in the atmosphere. The bigger fractal dimension particles have, the stronger adsorption of toxic particulate matter atmospheric particles have. Then, atmospheric particles will greatly affect human health.

The controlled particulate matter (**Figure 8**) can adsorb more toxic particulate matter and cause it to control the settlement and reduce the environmental pollution. In addition, if the particulate matter still cannot settle after control, it will be controlled as in **Figure 6** (fractal dimension is 2.029) and **Figure 7** (fractal dimension 2.021, 2.031 and 2.038, respectively) shape structure, in order to reduce the adsorption of particles on toxic particles and the harm to human health.

5. Conclusions

The analysis and its agglomeration control of soot particles fractal growth provides a new idea for the development of particulate matter traps and also provides a new solution for reducing environmental pollution. Based on the fractal growth physical model of soot particles from large diesel agriculture machinery, this chapter simulates the morphological structure of collision for the single particles and single particles, single particle and clusters, clusters and clusters, firstly. Moreover, combining with the collision frequency, the fractal growth is controlled to agglomeration using the main environmental factors interference for diesel engine soot particles, in order to make them condensed into regular geometry or larger density particles, reduce the viscous drag for capturing by the capturer or settlement and to realize the control of the pollution of the environment.

If the particles cannot settle, they can be controlled to reduce the adsorption of inhalable particles to toxic particles and reduce the harm to human health.

This chapter simulates the control of the aggregation fractal growth trend of diesel soot particles. The results of numerical simulation show that the proposed method is feasible and effective, which will help to understand and analyze the physical mechanism and kinetics of non-equilibrium condensation growth behavior of the actual carbon smoke particles and provide the solution to further reduce emissions of the inhalable particulate matter from diesel engines.

Acknowledgements

The work was supported by the National Natural Science Foundation of China (No. 31700644), Postdoctoral Science Foundation of China (Nos. 2015 M582122 and 2016 T90644), Key research and development project of Shandong Province(Nos. 2016ZDJS02A07 and 2017GNC12105). Agricultural machinery research and development project of Shandong Province (No. 2018YF004)." The outstanding youth talent cultivation plan" project of Shandong Agriculture University (No. 564032). The authors are grateful to all study participants. The authors declared that they have no conflicts of interest to this work. We declare that we do not have any commercial or associative interest that represents a conflict of interest in connection with the work submitted.

Nomenclature

α	the coefficient of equation.
$\eta(x,y)$	the condensation temperature.
c_n	the judgment factor.
$u(x,y)$	the initial value of the digitization.
x_s	the center of the ball B_m.

Author details

Ping Liu* and Chunying Wang
College of Mechanical and Electronic Engineering, Shandong Agricultural University, Taian, China

*Address all correspondence to: liupingsdau@126.com

IntechOpen

References

[1] Poran A, Tartakovsky L. Performance and emissions of a direct injection internal combustion engine devised for joint operation with a high-pressure thermochemical recuperation system. Energy. 2017;**124**:214-226

[2] Robelia B, Mcneill K, Wammer K, et al. Investigating the impact of adding an environmental focus to a developmental chemistry course. Journal of Chemical Education. 2010; **87**(2):216-220

[3] Kittelson DB. Engines and nanoparticles: A review. Journal of Aerosol Science. 1998;**29**(5–6):575-588

[4] Tian H, Liao Z. Progress on the formation mechanism of biomass soot particles. Clean Coal Technology. 2017; **23**(3):7-15

[5] Shuai J et al. Review of formation mechanism and emission characteristics of particulate matter from automotive gasoline engines. Transactions of CSICE;**2016**(2):105-116

[6] Maozhao XIE. The Computational Combustion Theory of the Internal Combustion Engine. Dalian: Dalian Institute of Technology Press; 2005

[7] M Balthasar, M Frenklach. Monte-Carlo simulation of soot particle coagulation and aggregation: The effect of a realistic size distribution. Proceedings of the Combustion. 2005; **30**(1):1467-1475

[8] Zhang L, Liu ST. Directed control for fractal growth with environmental disturbance. Control Theory & Applications. 2011;**28**(12):1786-1790

[9] Liu P, Liu ST. Nonlinear generalized synchronization of two different spatial Julia sets. Control Theory & Applications. 2013;**30**(9):1159-1164

[10] Hu E, Hu X, Liu T, et al. The role of soot particles in the tribological behavior of engine lubricating oils. Wear. 2013;**304**(1–2):152-161

[11] Liu Y, Tao F, Foster DE, et al. Application of a multiple-step phenomenological soot model to HSDT diesel multiple injection modeling. In: 2005 SAE World Congress. Warrendale: SAE Transactions; 2005. pp. 1141-1156

[12] Pang KM, Ng HK, Gan S. Investigation of fuel injection pattern on soot formation and oxidation processes in a light-duty diesel engine using integrated CFD-reduced chemistry. Fuel. 2012;**96**(7):404-418

[13] Ming-rui W, Hui-ya Z, Liang K, Wei-dong Z. Numerical simulation for the growth of diesel particulate matter and its influential factor analysis. Advances in Natural Science. 2008; **9**(18):1028-1033

[14] Koto F, Yanagimoto T, Mori K, et al. The Clarification of Fuel-Vapor Concentration on the Process of Initial Combustion and Soot Formation in a Di Diesel Engine. 2017;**2003**.1:1-253-1-258

[15] A Witten T, Sander LM. Diffusion-limited aggregation, a kinetic critical phenomenon. Physical Review Letters. 1981;**47**(19):1400 (p 4)

[16] Jizhong Z. Fractal. Beijing: Tsinghua University Press; 2011

[17] Deng X, Dav RN. Breakage of fractal agglomerates. Chemical Engineering Science. 2017;**161**:117-126

[18] Pfeifer P, Avnir D. Chemistry in noninteger dimensions between two and three. I. Fractal theory of heterogeneous surfaces. The Journal of Chemical Physics. 1983;**79**(7):3558-3565

[19] Sun J, Qiao W, Liu S. Controlling fractal diffusion of differently-sized soot particles. International Journal of Bifurcation and Chaos. 2018

[20] Chou CK, Lee CT. On the aerodynamic behavior of fractal agglomerates. Journal of Aerosol Science. 1997;**28**(2):620-635

Chapter 3

Kinetic Modeling of Photodegradation of Water-Soluble Polymers in Batch Photochemical Reactor

Dina Hamad, Mehrab Mehrvar and Ramdhane Dhib

Abstract

Synthetic water-soluble polymers, well-known refractory pollutants, are abundant in wastewater effluents since they are extensively used in industry in a wide range of applications. These polymers can be effectively degraded by advanced oxidation processes (AOPs). This entry thoroughly covers the development of the photochemical kinetic model of the polyvinyl alcohol (PVA) degradation in UV/H_2O_2 advanced oxidation batch process that describes the disintegration of the polymer chains in which the statistical moment approach is considered. The reaction mechanism used to describe the photo-degradation of polymers comprises photolysis, polymer chain scission, and mineralization reactions. The impact of operating conditions on the process performance is evaluated. Characterization of the polymer average molecular weights, total organic carbon, and hydrogen peroxide concentrations as essential factors in developing a reliable photochemical model of the UV/H_2O_2 process is discussed. The statistical moment approach is applied to model the molar population balance of live and dead polymer chains taking into account the probabilistic chain scissions of the polymer. The photochemical kinetic model provides a comprehensive understanding of the impact of the design and operational variables.

Keywords: kinetic modeling, population balance, free radical-induced degradation, advanced oxidation process, water-soluble polymer

1. Introduction

The growing turnover and consumption of synthetic water-soluble polymers generate a huge amount of wastes during production, use, and disposal off. After usage, water-soluble polymers are expected to end up in rivers, lakes, oceans and even in wastewater treatment plants, thus creating a potential pollution hazard. In contrast to biopolymers, water-soluble polymers are resistant to microorganisms-mediated biodegradation [1–3]. Synthetic water-soluble polymers cover a wide range of highly varied families of products and have numerous interesting applications.

One of the concerns is the accumulation of such non-biodegradable water-soluble polymers in the environment. Particularly, polyvinyl alcohol (PVA) is one of the most commercially important water-soluble synthetic polymers with an annual worldwide production of 650,000 tons. PVA polymers are abundant in wastewater effluents due to the extensive usage in paper and textile industries that accordingly generates significant amounts of PVA in wastewater streams [4, 5]. The PVA polymers are used in industry as paper and textile coatings, and also as laundry packing materials [6]. Its iodine complexes are widely used as polarization layers in liquid crystal displays (LCDs) [7]. As the production of PVA finds new markets, its consumption grows and the volume of wastewater containing PVA increases during its production and consumption. Moreover, PVA is highly soluble in water, and it leaches readily from soil into groundwater creating environmental issues. PVA polymers act as collector reagents that can be either chemisorbed or physically adsorbed since these polymer compounds have oxygen hetero-atoms capable of binding to different metal ions effectively and increase the mobilization of heavy metals from sediments of lakes and oceans which results in accumulation of hazardous materials. [6, 8–11]. Besides, The PVA solutions exhibit high surface activity supporting the formation of foams which can hinder the transport of oxygen into water streams. Therefore, the removal of PVA from wastewater systems is essential.

Conventional biological technologies are not efficient to breakdown PVA polymer chains since the degradation capacity of most microorganisms towards PVA is very limited and requires specially adapted bacteria strains [1]. In addition, wastewaters containing PVA can cause foam formation in biological equipment which inhibits the activity of aerobic microorganisms due to oxygen absence that results in unstable operation with low performance [9]. As a result, the advanced oxidation processes are utilized as alternative treatment techniques for the treatment of polymeric wastewater systems. The advanced oxidation technologies are proven to be effective in treating industrial wastewater [10, 11]. AOPs involve the formation of strong oxidants such as hydroxyl radicals and the reaction of these oxidants with pollutants in wastewater. In wastewater treatment applications, AOPs usually refer to a specific subset of processes that involve H_2O_2, O_3, and UV light as shown in the schematic diagram in **Figure 1**.

The degradation of water-soluble polymers by different AOPs is studied in the open literature whether those polymers are refractory, toxic, hazardous or recalcitrant compounds. Recent studies on the removal of PVA have focused on radiation-induced oxidation process such as photo-Fenton [7], photocatalytic processes [7, 12], radiation-induced electrochemical process [13], and UV/H_2O_2 process [14–17]. Even though the degradation of a polymer component must be assessed by the reduction and analysis of its molecular weights, there are only a few studies in the open literature on the devolution of the molecular weight size distributions of water-soluble polymers [9, 17]. Also, the residual hydrogen peroxide is still a challenging issue in the UV/H_2O_2 process which has been overlooked in some studies.

Furthermore, there is little information on the photochemical mechanism of the photo-oxidative degradation of PVA polymer solutions in a UV/H_2O_2 process. Recently, several attempts have been made to comprehend the chemical kinetics dominating thermal degradation of water-soluble polymers and assuming constant pH [18, 19]. Besides, no data is available on the distribution of the molecular weights of the polymer being degraded.

Under UV radiation, polymer chains are broken down into oligomers (short-chain polymers), dimers and monomers. Enhanced photo-degradation of polymer can lead to a broader distribution of molecular weights, indicating that the degraded polymer becomes less and less uniform. This behavior is expected for degraded polymers, as irradiation promotes an increase in the number of polymer chains, lowering the molecular weight, and consequently increasing the polydispersity. Hence, polymer degradation is a fragmentation process in which population balance

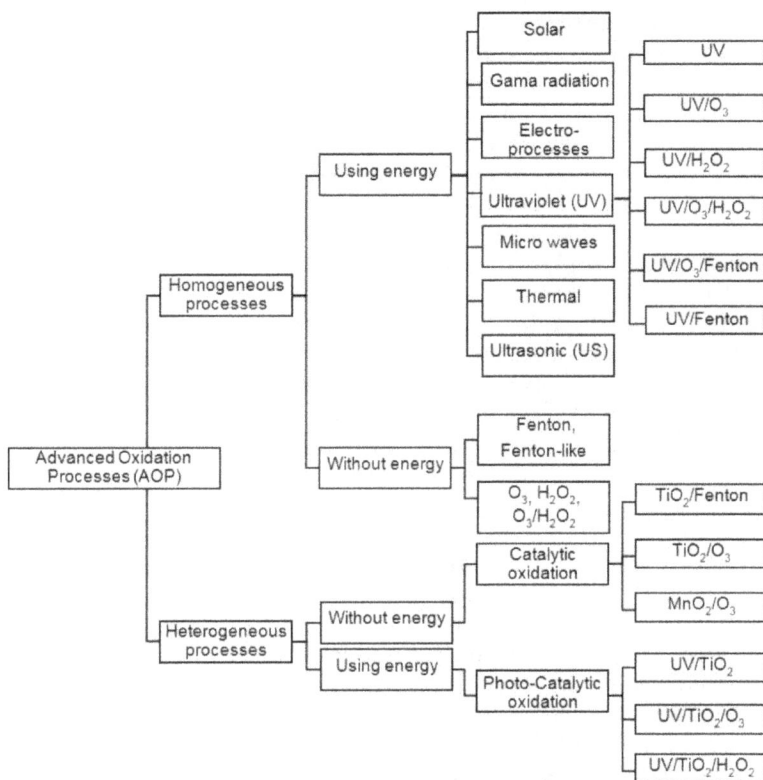

Figure 1.
Schematic classification of the advanced oxidation processes.

concepts is often applied in fragmentation models to describe how the distributions of different size entities evolve over the time of reaction.

The degradation of high molecular weight polydisperse materials results in the formation of a large number of polymeric chains with different chain lengths and various chemical compositions, i.e., the number of branches. Population balance based models have been developed to study the molecular weight decrease of polymers in a fragmentation process [5, 13, 20–22]. Population balance approach is generally employed to model the size distribution of the macromolecular compound during polymerization, depolymerization, and chain breakage. The population balance model is a balance equation of species of different sizes, and it is similar to the mass, energy, and momentum balances, to track the changes in the size distribution.

Few important studies have been done to understand the chemical kinetics that dominates the degradation of water-soluble polymers with UV radiation. Even though encouraging results on the degradation of polymers were obtained, data on the molecular weight distribution of the treated polymer need to be collected. Hence, it is worthwhile to investigate further the degradation process of synthetic polymer and the devolution of their molecular weight distributions. Other studies have theoretically analyzed the thermal degradation of synthetic polymers and provided a mathematical interpretation of the polymer chain scissions. The photochemical mechanism and kinetic modeling of the photo-oxidative degradation of water-soluble polymers have been investigated in several studies. Nevertheless, the proposed mechanisms may be complex and not well-established. The majority of mathematical approaches to polymer degradation consider only the polymer molecular-weight distribution (MWD) or chain-length distribution. The treatment of the wastewater streams contaminated with PVA polymers is studied using different processes [7, 12, 13, 16]. The kinetics models proposed in these studies were validated using total organic carbon (TOC) data instead of polymer concentrations or polymer molecular weights, and a constant pH was assumed.

The photo-oxidative degradation of water-soluble polymers in laboratory scale photochemical reactors is the focus of this chapter. The photochemical kinetic model of the polymer degradation in UV/H_2O_2 process that describes the polymer chain scission is discussed in which the statistical moment approach is presented. The development of a photochemical kinetic model incorporates the population balance of all chemical species. Considering the probabilistic nature of the polymer fragments, the statistical moment approach is applied for modeling the population balance of live and dead polymer chains, which allows estimating the polymer average molecular weights as a function of radiation time. The model also considered the effect of process parameters on the decrease of polymer molecular weight, hydrogen peroxide residual, and the acidity of the treated solution.

2. UV/H_2O_2 system description

The critical design parameters in the UV/H_2O_2 process include the H_2O_2 dose, the UV lamp type and intensity, and the reactor contact time. Basic UV reactor design configurations used for the removal of polymers from wastewater depend mainly on the flow rate. The tower design is typically utilized for large-scale applications. In the tower configuration, multiple UV lamps are arranged horizontally within a single large reactor vessel with the contaminated water flowing perpendicularly past the UV lamps [23]. For small-scale systems, a single UV lamp per reactor vessel is arranged vertically. For example, a small-scale system may consist of three individual reactor vessels in series, each containing one UV lamp in a vertical position.

A typical laboratory-scale batch recirculation UV/H_2O_2 system consists of an annular photoreactor, a large volume reservoir tank, centrifugal pump, and heat exchanger. The circulation tank contains the polymer solution for treatment. The hydrogen peroxide is injected into the circulation tank. A centrifugal magnet pump is placed on the circulation line to maintain a steady flow of the aqueous polymer solution between the tank and photoreactor. A flow meter is used to measure the circulation rate. The cylindrical photoreactor is made of steel vessel of annual shape and is connected to the circulation tank. The reactor is equipped with an internal quartz glass in which a low-pressure mercury UV lamp is mounted at its centerline of the cylinder with stainless steel housing. The annular photoreactor should have a very small annular space to assure a uniform light distribution in the photoreactor. Most AOPs are modular processes. Therefore, more than one reactor can be employed in series mode to obtain higher retention times or in parallel mode to process larger volumes to achieve the desired effluent for a given flow rate.

3. Characterization of polymeric wastewater

Determination of the polymer molecular weight, TOC content, and residual hydrogen peroxide are crucial parameters to assess the performance of the photodegradation process. The treated samples are analyzed using gel permeation chromatography (GPC) to determine the molecular weights of the degraded polymer samples. The GPC theory depends on the principle of size exclusion; therefore, when a polymer solution is passed through a column of porous particles, large molecules cannot enter the pores of the packing and hence, they elute first. However, smaller molecules that can penetrate or diffuse into the pores are retained for a while in the column and then elute at a later time. Thus, a sample is fractionated by molecular hydrodynamic volume, and the resulting profile describes the molecular weight distribution. A concentration detector (e.g., differential refractometer (RI) or UV detector) is placed downstream of the columns to measure the concentration of each fraction as a function of time. The actual method for determining the molecular weight averages and the MWD depends upon the attached detectors. GPC provides a convenient, quick, and relatively easy method which can be used on a routine basis for determining the various moments of molecular weight.

The extent of degradation reactions to CO_2 is monitored by measuring the total organic carbon content of the samples. TOC analyzer is based on the oxidation of organic compounds to carbon dioxide and water, with subsequent quantities of carbon dioxide. The TOC analyzer subtracts the inorganic carbon (CO and CO_2) and reports the total organic carbon, which is a close approximation of organic content. The amount of carbon dioxide generated upon oxidation of the organic carbon in the sample was determined by the non-dispersive infra-red (NDIR) detector, which is sensitive to low levels of TOC.

The reduction of hydrogen peroxide concentration during the degradation reaction is determined using spectrophotometer method using 9-dimethyl-1, 10-phenanthroline (DMP) method. The most common way of measuring hydrogen peroxide residual in wastewater is DMP-spectrophotometer method. 9-Dimethyl-1,10-phenanthroline (DMP) method is based on the chemical reduction of copper (II) by hydrogen peroxide in the presence of DMP, thus forming a bright yellow (copper (II) – DMP) complex that is directly determined by UV–vis spectrophotometer [24]. The DMP method appears to be simple, robust, and rather insensitive to interference. Intermediate compounds such as acetic and formic acids, formaldehyde, and acetaldehyde, which are formed by the decomposition of organic matter exposed to AOPs, do not interfere with the DMP method.

4. Polymer chain scission mechanism

The degradation of polymers by advanced oxidation processes is mainly due to free-radical-induced chain scission that led to successive oxidation reactions which result in lower molecular weight polymer fragments. The chain scission reaction is a chemical reaction between the macromolecular compounds (polymers) and end/mid-chain radicals. As the reaction progresses, the large polymer molecules eventually break down into live and dead polymer chains of lower molecular weights. A further molecular disintegration can ultimately lead to carbon dioxide and water as final products in case of complete mineralization.

In other words, the chain scission reaction can be defined as a bond scission that takes place in the backbone of the polymer chain. The chain scission reaction increases the number of polymer chains and reduces the polymer molecular weight [25]. Consequently, the chain scission results in an increase in the polydispersity of the polymer sample which represents the breadth of the distribution curve.

The concept of polymer degradation may be explained by chain-end scission or random chain scission mechanisms where chain breaking occurs at a random location along the chain. Therefore the molecular weight decreases continuously with the extent of reaction. Chain-end scission is considered as a special case of random chain scission where the chain scission reactions are occurring most likely at the polymer chain end that results in a release of a single monomer molecule. Random chain scission is the reverse of step-growth polymerization while chain-end scission is the reverse of chain growth polymerization [26]. Aarthi et al. [14] studied the photodegradation of water-soluble polymers by combined ultrasonic and ultraviolet radiation and found that the degradation process is controlled by random and midpoint scission. On the other hand, Konaganti and Madras [27] investigated the photocatalytic degradation of polymethyl methacrylate, polybutyl acrylate, and their copolymers by random and chain-end scissions.

In the photodegradation of PVA polymer, the random chain scission mechanism initially dominates which experimentally proved by the rapid decrease in the polymer molecular weights. In random chain scission, all bonds may have an equal probability of being cleaved along the polymer chain. Apparently, the degradation process leads to a steep decrease in molecular weights. The chain cleavage occurs and effectively shortens the polymer chains [17]. It can be concluded that PVA degradation occurs mostly by random chain scission which explains the drastic decrease in the polymer concentration.

4.1 Polymer average molecular weights

Polymer molecules are made of repeat monomer units that chemically bonded into long chains. The chain length is often expressed in terms of the molecular weight of the polymer chain, related to the relative molecular weight of the monomer and the number of monomer units connected in the chain.

The molecular weight of a polymer is described by the average values of the molecular weights of the polymer chains. The molecular weight distribution (MWD) is the distribution of sizes in a polymer sample while the polydispersity index (PDI) represents the breadth of the distribution curve. Thus, the polydispersity index is used as a measure of the broadness of molecular weight distribution of a polymer sample. Most synthetic water-soluble polymers are polydisperse since they contain polymer chains of unequal lengths. The increase in the polydispersity index results in broader molecular weight distribution. The PDI is defined as the ratio of weight average molecular weight (M_w) to the number average molecular weights (M_n). The molecular weight of a polymer is not a single value since polymer molecules even those of the same type, have different sizes, so the method of averaging mainly determines the average molecular weight. The number average molecular weight is considered as the ordinary arithmetic average of the molecular weights of the polymer while the weight average molecular weight is determined by measuring the weight of each species in the sample, rather than the number of molecules of each size.

Enhanced photodegradation of polymer by UV radiation can lead to a wider distribution of molecular weights because the polymer chains are broken down into short-chain polymers such as oligomers, dimers, and monomers [28]. The irradiation promotes the decrease in the polymer molecular weights and the increase in the polydispersity of the molecular weight distribution of the degraded polymer as shown in **Figure 2**.

The shape of the molecular weight distribution changes as a function of the treatment time. The untreated PVA has a uniform narrow distribution with a polydispersity index (PDI) close to unity. During the degradation process, the distribution shifts to the left as the polymer molecular weight was considerably lowered. Song and Hyuan [29] confirmed the shifting of MWD and the generation of monomer by chain-end scission at the thermal degradation of polystyrene in a batch reactor. The broadness of the molecular weight distribution which is expressed by an increase in polydispersity is due to the fragmentation and chain-scission mechanism of the polymer degradation during the UV/H_2O_2 process.

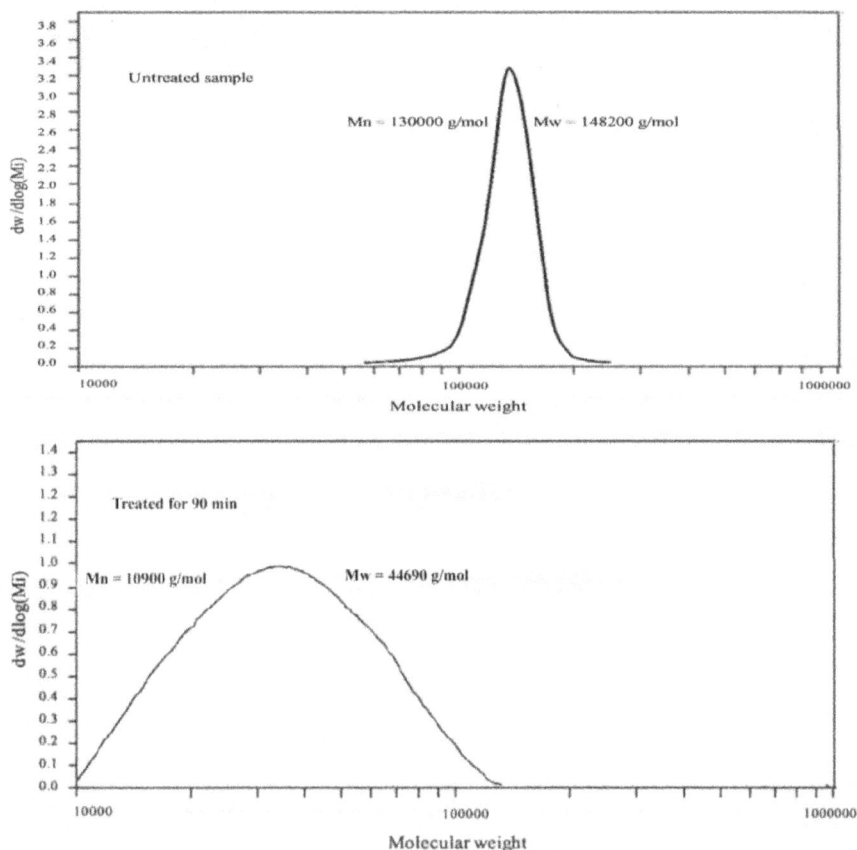

Figure 2.
Effect of radiation time on MWD of the degraded PVA polymer by UV/H₂O₂ process (data from [17]).

5. Photodegradation kinetic model development

The principle in the AOP process is the formation of hydroxyl radicals which react immediately with organic contaminants in the wastewater streams. The hydroxyl radicals are highly reactive because of the presence of unpaired electrons. Oxidation reactions that produce radicals tend to be followed by additional oxidation reactions between the radical oxidants and the intermediate products until thermodynamically stable oxidation products are formed at complete mineralization of the pollutant.

Usually, the mineralization starts directly with pollutant degradation, however, for PVA polymers it occurs at a later stage of the reaction. In this case, it is desired to model a specific polymer degradation as the TOC is not the right parameter to choose for the development of an adequate model for polymer disintegration in a photo-oxidation process. It is plausible to develop a model that takes into account the polymer molecular weights.

Under the effect of UV light of a specific wavelength and using an oxidant such as hydrogen peroxide, water-soluble polymer chains can break down into smaller chains. Under the effect of radiation energy, chemical bonds of polymer chains are destabilized and weakened. The chain scission reaction is, therefore, initiated and it is defined as a bond scission that takes place in the backbone of the polymer chain.

As the reaction progresses, the large polymer molecules P_r eventually break down into live and dead polymer chains of lower molecular weights, and consequently, new intermediate polymeric components are formed. A further molecular disintegration can ultimately lead to carbon dioxide and water as final products in case of complete mineralization according to the following reaction:

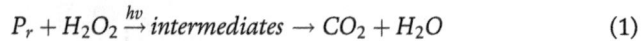

$$P_r + H_2O_2 \xrightarrow{hv} intermediates \rightarrow CO_2 + H_2O \tag{1}$$

Under the UV irradiation, the photolysis of hydrogen peroxide generates hydroxyl radicals as follows:

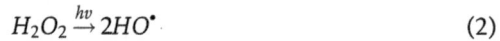

$$H_2O_2 \xrightarrow{hv} 2HO^\bullet \tag{2}$$

The highly reactive hydroxyl radical can undergo a series of promoted dissociation reactions. Several authors [30, 33, 40, 41] have proposed a detailed chemical kinetic mechanism of hydrogen peroxide decomposition. Photolysis reactions of hydrogen peroxide (Reactions (3)–(15)) and the rate constants are provided in **Table 1**.

No.	Reaction mechanism	Rate constant		Reference
(3)	$HO^\bullet + H_2O_2 \xrightarrow{k_1} HO_2^\bullet + H_2O$	2.7×10^7	L/mol s	[30]
(4)	$H_2O_2 + HO_2^\bullet \xrightarrow{k_2} HO^\bullet + H_2O + O_2$	3.0	L/mol s	[31]
(5)	$H_2O_2 + O_2^{\bullet-} \xrightarrow{k_3} HO^\bullet + OH^- + O_2$	13×10^{-2}	L/mol s	[32]
(6)	$HO^\bullet + HO_2^- \xrightarrow{k_4} HO_2^\bullet + OH^-$	7.5×10^9	L/mol s	[33]
(7)	$O_2^{\bullet-} + H^+ \xrightarrow{k_5} HO_2^\bullet$	1.0×10^{10}	L/mol s	[34]
(8)	$HO^\bullet + HO^\bullet \xrightarrow{k_6} H_2O_2$	5.5×10^9	L/mol s	[30]
(9)	$HO_2^\bullet + HO^\bullet \xrightarrow{k_7} H_2O + O_2$	6.6×10^9	L/mol s	[35]
(10)	$HO_2^\bullet + HO_2^\bullet \xrightarrow{k_8} H_2O_2 + O_2$	8.3×10^5	L/mol s	[34]
(11)	$HO_2^\bullet \xrightarrow{k_9} O_2^{\bullet-} + H^+$	1.6×10^5	1/s	[34]
(12)	$HO_2^\bullet + O_2^{\bullet-} \xrightarrow{k_{10}} O_2 + HO_2^-$	9.7×10^7	L/mol s	[34]
(13)	$HO^\bullet + O_2^{\bullet-} \xrightarrow{k_{11}} O_2 + OH^-$	7.0×10^9	L/mol s	[36]
(14)	$H_2O_2 \xrightarrow{k_{12}} HO_2^- + H^+$	4.5×10^{-12}	1/s	[37]
(15)	$HO_2^- + H^+ \xrightarrow{k_{13}} H_2O_2$	2.0×10^{10}	L/mol s	[37]
(16)	$P_r + HO^\bullet \xrightarrow{k_{P_1}} P_r^\bullet + H_2O$	8.06×10^6	L/mol s	[38]
(17)	$P_r + HO_2^\bullet \xrightarrow{k_{P_2}} P_r^\bullet + H_2O_2$	4.69×10^{-1}	L/mol s	[38]
(18)	$P_r^\bullet \xrightarrow{k_p} P_{r-1}^\bullet + P_1$	3.66×10^2	1/s	[38]
(19)	$P_{r-s}^\bullet + P_s^\bullet \xrightarrow{k_{t_c}} P_r$	4.44×10^2	L/mol s	[38]
(20)	$P_1 + HO_2^\bullet + O_2^{\bullet-} \xrightarrow{k_{d_1}} 2HCOOH + HO^-$	1.89×10^6	L/mol s	[38]
(21)	$P_1 + HO_2^\bullet \xrightarrow{k_{d_2}} CH_3COOH + HO^\bullet$	1.35×10^2	L/mol s	[38]
(22)	$HCOOH \xrightarrow{K_{d_3}} HCOO^- + H^+$	1.77×10^{-4}		[39]

No.	Reaction mechanism	Rate constant		Reference
(23)	$CH_3COOH \overset{K_{a2}}{\rightleftharpoons} CH_3COO^- + H^+$	1.76×10^{-5}		[39]
(24)	$CH_3COOH + 2HO_2^{\bullet} \overset{k_{14}}{\rightleftharpoons} 2HCOOH + H_2O_2$	1.60×10^7	L/mol s	[39]
(25)	$HO^{\bullet} + CH_3COO^- \overset{k_{15}}{\rightarrow} {}^{\bullet}CH_2COO^- + H_2O$	3.20×10^9	L/mol s	[39]
(26)	$HCOOH + 2HO^{\bullet} \overset{k_{17}}{\rightarrow} CO_2 + 2H_2O$	1.30×10^8	L/mol s	[39]

Table 1.
Photolysis reactions of hydrogen peroxide and the rate constants.

The mechanism of degradation polymer solution using UV irradiation using hydrogen peroxide as an oxidant results in the generation of polymeric hydroxyl radicals, which undergo degradation reactions. The live polymer radicals $P_r^{\bullet}$ are the precursor of subsequent polymer chain breakage (Reactions (16)–(19)). The polymer radical may combine with another polymer radical to terminate the reaction (Reactions (21)). The scission products are radical and non-radical fragments (monomer or polymer with lower molecular weight) as follows:

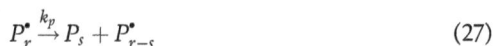

$$P_r^{\bullet} \overset{k_p}{\rightarrow} P_s + P_{r-s}^{\bullet} \qquad (27)$$

where $s = 1$ f0r chain-end scission and $2 \leq s \leq r$.

Reactions (20)–(26) represent the complete mineralization of polymer compounds. It has been experimentally proven that the acidity of the treated solution varies during the degradation reaction by the UV/H_2O_2 process [17, 39]. The pH decreases at the beginning of the reaction, and the solution becomes more acidic due to the formation of intermediate oxidation products such as carboxylic acids [42]. A regain in the pH of the solution is expected in case of complete mineralization as a result of the degradation of acidic compounds that are oxidized to carbon dioxide that escapes the system and water at the end of the reaction. The experimental findings indicate that there is evidence of the formation of acetic and formic acids associated with the degradation of the monomer (vinyl alcohol) produced at the complete degradation of PVA polymer. Therefore, the photochemical kinetic mechanism incorporates the acidity aspect of the solution as the polymer degradation progresses. The complete mineralization of polymer compounds and the production of by-products with no hazard to the environment (Reactions (20)–(26)) are considered as remarkable advantages of the advanced oxidation processes. A photochemical kinetic model was developed based on the mechanism presented in Reactions (1) to (27).

The polymer degradation reactions are assumed to be irreversible. Binary fragmentation is also considered to explain kinetics fragmentation in which a polymer of chain length r splits into two polymer units. The reaction rate constants are assumed independent of polymer chain length. The degradation reaction is carried out at a constant temperature with a good mixing condition. For a batch reactor with recirculation, negligible degradation of the polymer per pass and good mixing condition are assumed. Direct photolysis of the polymer without the presence of hydrogen peroxide is neglected.

The photochemical kinetic model describing the PVA polymer degradation by photo-oxidation comprises a radiation energy balance coupled with a molar balance of the chemical species participating in the degradation reactions of the polymer. The quantum yield of PVA is usually negligible since there is no measurable change in PVA molecular weight under UV radiation alone [17]. The molar absorptivity of PVA polymer is determined using spectrophotometer by measuring the absorbance of different concentrations of PVA aqueous solutions at a wavelength of 254 nm.

For the kinetics, the general molar balance equation (Eq. 28) [43] must be applied to the recirculating batch photoreactor.

$$\frac{\partial c_i}{\partial t} + \nabla.N_i = R_i \tag{28}$$

Assuming that the system works under the well-stirred conditions ($\nabla.N_i = 0$), the ratio of the photoreactor volume to the total volume $\ll 1$, and high recirculating flow rate to ensure small conversion per pass, the rate of the change of the concentration in the tank could be written as follows [44]

$$\frac{dC_i}{dt} = \frac{V_{ph}}{V_T} \sum_{j=1}^{m} R_{ij}, \quad C_i(0) = C_{i0} \tag{29}$$

In which $C_i(t)$ is the ith component concentration, V_{ph} is the volume of the photoreactor, V_T is the volume of the whole system, $C_i(0)$ is the initial molar concentration of species i, and R_{ij} is the chemical reaction rate of component i in reaction j (j = 1,2,.,m).

According to the basic photochemical mechanism given in **Table 1**, the mole balance of small molecule species gives the following reaction rate equations:

$$\frac{1}{\alpha}\frac{d[H_2O_2]}{dt} = -R_{UV,H_2O_2} - k_1[HO^\bullet][H_2O_2] - k_2[HO_2^\bullet][H_2O_2] - k_3[O_2^{\bullet-}][H_2O_2]$$
$$+ k_6[HO^\bullet]^2 + k_8[HO_2^\bullet]^2 + k_{p2}[HO_2^\bullet][P_r] - k_{12}[H_2O_2] + k_{13}[HO_2^-][H^+]$$
$$+ k_{14}[CH_3COOH][HO_2^\bullet] - k_{15}[HCOO^-][H_2O_2] \tag{30}$$

$$\frac{1}{\alpha}\frac{d[HO^\bullet]}{dt} = 2R_{UV,H_2O_2} - k_1[HO^\bullet][H_2O_2] + k_2[HO_2^\bullet][H_2O_2] + k_3[O_2^{\bullet-}][H_2O_2]$$
$$- k_4[HO_2^-][HO^\bullet] - 2k_6[HO^\bullet]^2 - k_7[HO^\bullet][HO_2^\bullet] - k_{p1}[HO^\bullet][P_r]$$
$$- k_{11}[HO^\bullet][O_2^{\bullet-}] - k_{d1}[HO^\bullet][P_1] + k_{d2}[HO_2^\bullet][P_1] - 2k_{16}[HO^\bullet]^2[HCOOH] \tag{31}$$

$$\frac{1}{\alpha}\frac{d[H^+]}{dt} = -k_5[O_2^{\bullet-}][H^+] + k_9[HO_2^\bullet] + k_{12}[H_2O_2] - k_{13}[H^+][HO_2^-] + k_{a1}[HCOOH]/$$
$$[HCOO^-] + k_{a2}[CH_3COOH]/[CH_3COO^-] \tag{32}$$

$$\frac{1}{\alpha}\frac{d[HCOOH]}{dt} = k_{d1}[HO^\bullet][P_1] + k_{a1}^{-1}[H^+][HCOO^-] + k_{14}[HO_2^\bullet]^2[CH_3COOH]$$
$$- k_{16}[HO^\bullet]^2[HCOOH] \tag{33}$$

$$\frac{1}{\alpha}\frac{d[CH_3COOH]}{dt} = k_{d2}[HO_2^\bullet][P_1] + k_{a2}^{-1}[H^+][CH_3COO^-] - k_{14}[HO^\bullet]^2[CH_3COOH] \tag{34}$$

$$\frac{1}{\alpha}\frac{d[HO_2^\bullet]}{dt} = k_1[HO^\bullet][H_2O_2] - k_2[HO_2^\bullet][H_2O_2] + k_4[HO_2^-][HO^\bullet] + k_5[O_2^{\bullet-}][H^+]$$
$$- k_7[HO^\bullet][HO_2^\bullet] - 2k_8[HO_2^\bullet]^2 - k_9[HO_2^\bullet] - k_{p2}[HO_2^\bullet][P_r]$$
$$- k_{10}[HO_2^\bullet][O_2^{\bullet-}] + k_{d1}[HO^\bullet][P_1] - k_{d2}[HO_2^\bullet][P_1] - 2k_{14}[HO_2^\bullet]^2[CH_3COOH] \tag{35}$$

$$\frac{1}{\alpha}\frac{d\left[HO_2^-\right]}{dt} = -k_4\left[HO_2^-\right]\left[HO^{\bullet}\right] + k_{10}\left[HO_2^{\bullet}\right]\left[O_2^{\bullet-}\right] + k_{12}[H_2O_2] - k_{13}\left[HO_2^-\right]\left[H^+\right]$$

$$(36)$$

$$\frac{1}{\alpha}\frac{d\left[O_2^{\bullet-}\right]}{dt} = -k_3\left[O_2^{\bullet-}\right][H_2O_2] - k_5\left[O_2^{\bullet-}\right]\left[H^+\right] + k_9\left[HO_2^{\bullet}\right] - k_{10}\left[HO_2^{\bullet}\right]\left[O_2^{\bullet-}\right]$$
$$- k_{11}\left[HO^{\bullet}\right]\left[O_2^{\bullet-}\right]$$

$$(37)$$

$$\frac{1}{\alpha}\frac{d[P_1]}{dt} = k_p p_r^{\bullet} - k_{d1}\left[HO_2^{\bullet}\right]\left[O_2^{\bullet-}\right][P_1] - k_{d2}\left[HO_2^{\bullet}\right][P_1]$$

$$(38)$$

$$R_{UV,i} = -\varnothing_i f_i I_o\left(-\exp\left(-2.303\,b\sum_{i=1}^{N}\varepsilon_i.C_i\right)\right)$$

$$(39)$$

where α is defined as the ratio of photoreactor volume V_{ph} to the total volume of the system V_T, $\varnothing_i$ is the number of moles of the pollutant transformed per number of photons of wavelength λ absorbed by the pollutant, b is the path length of the ray through the medium, ε is the molar absorptivity, f_i is the fraction of the UV irradiation absorbed by the ith chemical species, I_o is the incident light intensity emitted at the source, and the C_i is the ith species concentration.

The molar balance of the macromolecules P_r and $P_r^{\bullet}$ in Reactions (16) to (21) requires special modeling approach as the PVA polymer is randomly broken down, polymer chains species of different sizes are subsequently generated, and they are expected to degrade further. The concept of the population species is considered to express the variations of the photochemical degradation of PVA. The random degradation of polymer chains of length r can be described using breakage population balance of all polymer species.

Generally, the moment operation is introduced as an easier method to transform the integro-differential equations in the continuous kinetics model or the sum in the discrete model to ordinary differential equations. McCoy and Madras [45] and Stickle and Griggs [46] provided simple mathematical expressions for the discrete model. The macromolecular reactions show that the polymer consists of degrading active polymer radicals $P_r^{\bullet}$ and dead polymer P_r. Polymer degradation is described by a discrete approach so that a mass balance provides a difference-differential equations. The net accumulation rate of dead polymer chains of chain length r is given as follows [38]:

$$\frac{1}{\alpha}\frac{d\left[p_r\right]}{dt} = -R_{UV,PVA} - k_{p_1}\left[HO^{\bullet}\right]\left[p_r\right] - k_{p_2}\left[HO_2^{\bullet}\left[p_r\right] + k_p\sum_{s=1}^{r}\Omega(r,s)p_s^{\bullet} + k_{tc}\sum_{s=1}^{r}p_r^{\bullet}p_{r-s}^{\bullet}\right]$$

$$(40)$$

Similarly, the net accumulation rate of live polymer radicals of chain length r is expressed as:

$$\frac{1}{\alpha}\frac{d\left[p_r^{\bullet}\right]}{dt} = R_{UV,PVA} + k_{p_1}\left[HO^{\bullet}\right]\left[p_r\right] + k_{p_2}\left[HO_2^{\bullet}\right]\left[p_r\right] - k_p\left[p_r^{\bullet}\right] + k_p\sum_{s=1}^{r}\Omega(r,s)p_s^{\bullet}$$
$$- 2k_{tc}\sum_{s=1}^{r}p_s^{\bullet}p_{r-s}^{\bullet}$$

$$(41)$$

Using statistical mechanics, the concept of moments was applied to determine the molecular weight distribution of a polymer population. This reaction requires the production of a specified scission product from any of a range of macromolecules, so a stoichiometric kernel $\Omega(r,s)$ is employed for a polymer chain of length r to represent the probability of getting shorter polymer chain lengths r-s and s [47]. In general, polymer degradation occurs most likely by random chain scission. Therefore, it is postulated that there is a low probability of the occurrence of chain-end scission reactions. For random chain scission, the distribution of shorter polymer chains is given as follows [45, 46]:

$$\Omega(r,s) = 1/r \tag{42}$$

5.1 Polymer population balance

Polymer degradation is a fragmentation process in which population balance concepts is often applied in fragmentation models to describe how the distributions of different size entities evolve over the time of reaction. The degradation of high molecular weight polydisperse materials results in the formation of a large number of polymeric chains with different chain lengths and various chemical compositions. Population balance approach is generally employed to model the size distribution of the macromolecular compound during polymerization, polymer degradation, depolymerization, and chain breakage.

In 1971, Randolph and Larson [48] proposed a solution for the population balance equation (PBE) in a well-mixed batch system. They used the concept of moment transform to convert the population balance equations into ordinary differential equations. Population balance based models have been developed to study the molecular weight decrease of polymers in a fragmentation process by advanced oxidation processes [18, 20, 22, 49]. Microwave-assisted oxidative degradation as an emerging advanced oxidation technology was used for poly(alkyl methacrylate) degradation. Random chain scission and Continuous distribution kinetics were employed to determine the degradation rate of the polymer [50]. Photocatalytic degradation of polyacrylamide co-acrylic acid by random chain scission has been investigated by Vinu and Madras [51]. The rate coefficients were determined as a linear function of the composition of co-monomer. Madras and McCoy [52] studied the kinetics of oxidative degradation of polystyrene by di-tert-butyl peroxide provided the ratio of the rate parameters for both oxidizer and polymer decomposition by moment analysis assuming random chain scission mechanism. Population balance and moment equations are solved for rate parameters [21, 53]. The model proposed by McCoy and Wang [21] is sufficiently applicable to a variety of degradation processes. Moment equations can be applied in batch and continuous stirred tank reactor (CSTR) reactors for binary or ternary fragmentation.

The population balance model is a balance equation of species of different sizes, and it is similar to the mass, energy, and momentum balances, to track the changes in the size distribution. The benefit of the population models is that they provide a straightforward technique to derive expressions for the moments of the polymer distributions during the degradation reaction. Hulburt and Katz [54] applied the concept of moments to determine the molecular weight distribution of a polymer population for a dead and live polymer moments as follow:

$$\mu_n = \sum_{r=1}^{\infty} r^n p_r \tag{43}$$

$$\lambda_n = \sum_{r=1}^{\infty} r^n p_r^* \tag{44}$$

where p_r and p_r^* are the polymer and the polymer radical concentrations with chain length r, μ_n and λ_n are the nth moment of the quantities p_r and p_r^* and n having values of 0, 1, or 2 stands for zeroth, first, and second moments, respectively. The application of the moment method allows converting the discrete differential population balance equations into ordinary differential ones. The moments of dead polymer P_r and live polymer radical $P_r^\bullet$ are used to determine the average molecular weights of the polymer. Applying the statistical moment concept to Eqs. (43, 44) gives the following model of dead and live polymer moments for n = 0, 1, and 2, respectively:

$$\frac{1}{\alpha}\frac{d[\mu_0]}{dt} = -R_{UV,PVA} - k_{p1}[HO^\bullet]\mu_0 - k_{p2}[HO_2^\bullet]\mu_0 + k_{tc}\lambda_0^2 + k_p\lambda_0 \tag{45}$$

$$\frac{1}{\alpha}\frac{d[\mu_1]}{dt} = -R_{UV,PVA} - k_{p1}[HO^\bullet]\mu_1 - k_{p2}[HO_2^\bullet]\mu_1 + 1/2k_p\lambda_1 + k_{tc}\lambda_1\lambda_1 \tag{46}$$

$$\frac{1}{\alpha}\frac{d[\mu_2]}{dt} = -R_{UV,PVA} - k_{p1}[HO^\bullet]\mu_2 - k_{p2}[HO_2^\bullet]\mu_2 + 1/3k_p\lambda_2 + k_{tc}\lambda_2\lambda_2 \tag{47}$$

$$\frac{1}{\alpha}\frac{d[\lambda_0]}{dt} = R_{UV,PVA} + k_{p1}[HO^\bullet]\mu_0 + k_{p2}[HO_2^\bullet]\mu_0 - 2k_{tc}\lambda_0^2 \tag{48}$$

$$\frac{1}{\alpha}\frac{d[\lambda_1]}{dt} = R_{UV,PVA} + k_{p1}[HO^\bullet]\mu_1 + k_{p2}[HO_2^\bullet]\mu_1 - 1/2k_p\lambda_1 - 2k_{tc}\lambda_0\lambda_1 \tag{49}$$

$$\frac{1}{\alpha}\frac{d[\lambda_2]}{dt} = R_{UV,PVA} + k_{p1}[HO^\bullet]\mu_2 + k_{p2}[HO_2^\bullet]\mu_2 - 2/3k_p\lambda_2 - 2k_{tc}\lambda_0\lambda_2 \tag{50}$$

Using statistical mechanics, the concept of moments is applied to determine the molecular weight distribution of a polymer population. The number average molecular weight M_n and the weight average molecular weight M_w are calculated according to:

$$M_n = NACL . M_o \tag{51}$$

$$M_w = WACL . M_o \tag{52}$$

where M_o is the molecular weight of the monomer unit. The number average chain length (NACL) and the weight average chain length (WACL) are given by:

$$NACL = \frac{\mu_1 + \lambda_1}{\mu_0 + \lambda_0} \tag{53}$$

$$WACL = \frac{\mu_2 + \lambda_2}{\mu_1 + \lambda_1} \tag{54}$$

A parameter estimation scheme is typically performed for the polymer photodegradation model equations to estimate the rate constants that are not available in the open literature. The objective function is the summation of squared errors between the model predictions and experimental data for selected process variables. The parameter estimation scheme is formulated to determine the estimates of the rate constants by minimizing the objective function which is subjected to the kinetic model equations.

The validity of the kinetic model is examined by direct comparison of model predictions with experimental data of the process parameters such as polymer molecular weights, polymer concentration, hydrogen peroxide residual, and pH of the solution. The goodness-of-fit between experimental y_{exp} and predicted y_m data for each variable are then determined by calculating the root mean square error (RMSE) for n' data points. The good agreement between the model predictions and the experimental results confirms the adequacy of the developed photochemical kinetic model.

5.2 Model predictions of the process variables

The polymer average molecular weights decrease with irradiation time due to the chain cleavage that effectively shortens the polymer chains which supports the success of the degradation process. The profile of the polymer molecular weight with time during the degradation process can indicate the type and mechanism of the chain scission. For instance, the steep reduction in the molecular weights of the PVA polymer at the beginning of the degradation reaction under UV irradiation is caused by the random chain scission mechanism that dominates initially in the photo-oxidative degradation of polyvinyl alcohol. At the end of the degradation reaction, the chain scission reactions occur most likely at the polymer chain end releasing a single monomer molecule when the polymer has considerably degraded. Whereas, the PVA degradation occurs mostly by random chain scission at the beginning of the reaction which explains the drastic reduction in the polymer concentration as clearly shown in **Figure 3** for initial PVA concentration of 50 mg/L. For water-soluble polymers, it is common to use a different approach, based on discrete population balance equations, to model polymer degradation involving

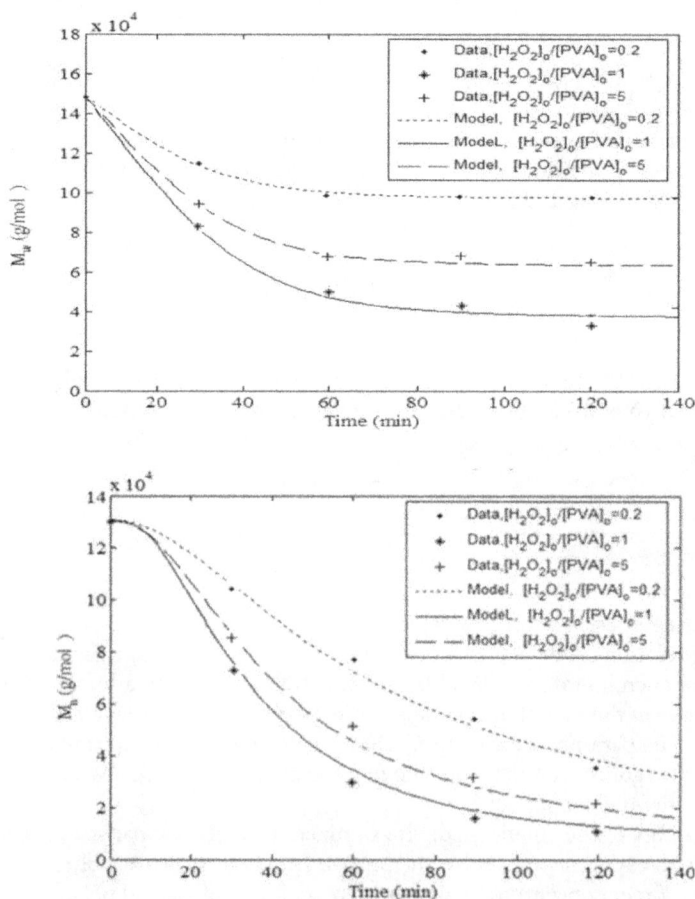

Figure 3.
Variation of the weight average M_w and number average M_n molecular weights of PVA at different $[H_2O_2]/[PVA]$ mass ratios in a batch UV/H_2O_2 photoreactor (data from [38]).

random scission and end-chain scission in order to predict the evolution of a population of molecules undergoing different scission mechanisms.

It is worth mentioning that hydrogen peroxide has a significant effect on the performance of the degradation process. The polymer molecular weight averages decrease with an increase of hydrogen peroxide concentration up to a certain limit. Therefore, a higher level of hydrogen peroxide has an adverse effect on the molecular weight reduction which can be interpreted by the scavenging effect of H_2O_2 over hydroxyl radicals which hinders the radical degradation since the amount of H_2O_2 added to the system is proportionally high [55]. The excess amount of hydrogen peroxide acts as a scavenger of hydroxyl radicals (Reaction (3) in **Table 1**) thus forming hydroperoxyl radicals. As shown earlier in the photochemical kinetics mechanism, the hydroperoxyl radical reacts with the PVA polymer (Reaction (17) in **Table 1**) [38]. Therefore, the probability of hydroxyl radicals attacking the polymer can be significantly reduced. The hydroperoxyl radicals are less reactive than hydroxyl radicals that subsequently suppress the degradation reaction. The photochemical model takes into account the scavenging effect of hydrogen peroxide by incorporating reaction rate equations of all radicals in order to enhance the reliability of the model.

The PVA polymers are effectively degraded in a UV/H_2O_2 photochemical reactor. In fact, the rates of polymer degradation and TOC removal did not match with each other. In fact, the TOC accounts for the carbon content of all chemical species, including PVA polymers. The difference between TOC and PVA removal efficiencies as shown in **Figure 4** is due to the presence of intermediate oxidation products and the non-degraded polymer residuals towards the end of reaction which can slightly increase the TOC content of the treated solution.

Figure 4 clearly illustrates the thresholds of the mass ratio of H_2O_2 and the polymer at which both the TOC removal and PVA degradation efficiencies at maximum values. In the advanced oxidation process, the amount of oxidant has to be experimentally determined according to the specified operating conditions for each pollutant so that the photochemical reaction performs at its best. Using excess hydrogen peroxide in the treatment process not only impedes the removal rate of the organic pollutants but also increase the hydrogen peroxide residual in the treated solution which can negatively affect the operating cost of the photoreactor system.

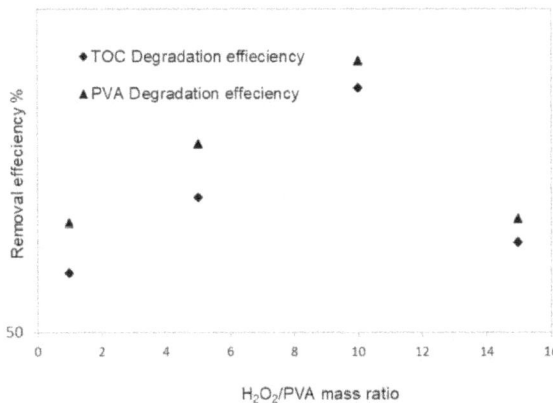

Figure 4.
PVA degradation and TOC removal efficiency for PVA (500 mg/L) degradation in UV/ H_2O_2 photoreactor [data from [17]].

6. Conclusions

The performance of the UV/H$_2$O$_2$ advanced oxidation process was evaluated for the degradation of polymeric wastewater in the batch photoreactor. The UV/H$_2$O$_2$ process can significantly modify the structure of the PVA polymer and be a potential practice for the degradation of water-soluble polymers in wastewater. Under the effect of UV light, hydrogen peroxide is readily decomposed into hydroxyl radicals of high reactivity which become oxidizing agents and can immediately attack the chains resulting in polymer disintegration.

A theoretical description of the UV/H$_2$O$_2$ process incorporates a population balance of polymer system and a molar balance of all chemical species to adequately represent the degradation of PVA polymer in a UV/H$_2$O$_2$ batch recirculating process. Modeling the photochemical degradation of the polymers represents a new approach to investigate the variations in polymer molecular weights. Considering the importance of oxidant in the advanced oxidation process performance, the dosage of hydrogen peroxide has to be experimentally determined for each polymer in order to achieve a better photochemical degradation of water-soluble polymers in wastewater. Incorporating the scavenging effect of hydrogen peroxide and the variation of the solution acidity is essential for the predictive quality and reliability of the photochemical model for degradation of polymers by UV/H$_2$O$_2$ process.

The photochemical mechanism and the photochemical kinetic model provide a framework for understanding the real characterization of the UV/H$_2$O$_2$ process and contribute to enhancing the design of industrial UV/H$_2$O$_2$ processes for the treatment of wastewaters contaminated with water-soluble polymers.

Acknowledgements

The authors would like to thank the editors for their efforts in improving the quality of the manuscript. The financial support of Ryerson University and the Natural Sciences and Engineering Research Council of Canada (NSERC) is greatly appreciated.

Nomenclature

C	molar concentration, mol/L
i	number of species
kp	rate constant of propagation, 1/s
kp$_1$	rate constant of propagation, L/mol s
kp$_2$	rate constant of propagation, L/mol s
kt$_c$	rate constant of termination by coupling, L/mol s
M$_n$	number average molecular weight of the polymer, g/mol
M$_w$	weight average molecular weight of the polymer, g/mol
P$_1$	monomer
P_r	dead polymer of chain length r
P_{r-s}	dead polymer of chain length r-s, where $1 \leq s < r$
$P_r^{\bullet}$	live radical of chain length r
$P_{r-s}^{\bullet}$	live radical of chain length r-s, where $1 \leq s < r$
R	rate of reaction, mol/L s
AOP	advanced oxidation process
DMP	9-dimethyl-1,10-phenanthroline

GPC	gel permeation chromatography
MWD	molecular weight distribution
NDIR	non-dispersive infra-red
PBE	population balance equation
PDI	polydispersity index
PVA	polyvinyl alcohol
TOC	total organic carbon
US	ultrasound
UV	ultraviolet

Author details

Dina Hamad[1*], Mehrab Mehrvar[2] and Ramdhane Dhib[2]

1 Chemical Engineering and Pilot Plant Department, National Research Center, Cairo, Egypt

2 Department of Chemical Engineering, Ryerson University, Toronto, Ontario, Canada

*Address all correspondence to: dhamad@ryerson.ca

IntechOpen

References

[1] Vitale P, Ramos P, Colasurdo V, Fernandez M, Eyler Y. Treatment of real wastewater from the graphic industry using advanced oxidation technologies: Degradation models and feasibility analysis. Journal of Cleaner Production. 2018. DOI: 10.1016/j.jclepro.2018.09.105

[2] Shonberger H, Baumann A, Keller W. Study of microbial degradation of polyvinyl alcohol (PVA) in wastewater treatment plants. American Dyestuff Reporter. 1997;**86**:9-18

[3] Solaro A, Corti A, Chillini E. Biodegradation of polyvinyl alcohol with different molecular weights and degree of hydrolysis. Polymers for Advanced Technologies. 2000;**11**:873-878

[4] Hamad D, Dhib R, Mehrvar M. Photochemical degradation of aqueous polyvinyl alcohol in a continuous UV/H_2O_2 process: Experimental and statistical analysis. Journal of Polymers and the Environment. 2016;**24**:72-83

[5] Ghafoori S, Mehrvar M, Chan P. Free-radical-induced degradation of aqueous polyethylene oxide by UV/H_2O_2: Experimental design, reaction mechanisms, and kinetic modeling. Industrial & Engineering Chemistry Research. 2012;**51**:14980-14993

[6] Sun W, Tian J, Chen L, He S, Wang J. Improvement of biodegradability of PVA-containing wastewater by ionizing radiation pre-treatment. Environmental Science and Pollution Research. 2012;**19**:3178-3184

[7] Zhang Y, Rong W, Fu Y, Ma X. Photocatalytic degradation of polyvinyl alcohol on Pt/TiO_2 with Fenton reagent. Journal of Polymers and the Environment. 2011;**19**:966-970

[8] Ciner F, Akal Solmaz SK, Yonar T, Ustun GE. Treatability studies on wastewater from textile dyeing factories in Bursa, Turkey. International Journal of Environment and Pollution. 2003;**19**:403-407

[9] Zhang SJ, Yu HQ. Radiation-induced degradation of polyvinyl alcohol in aqueous solutions. Water Research. 2004;**38**:309-316

[10] Rosario-Ortiz FL, Wert EC, Snyder SA. Evaluation of UV/H2O2 treatment for the oxidation of pharmaceuticals in wastewater. Water Research. 2010;**44**:1440-1448

[11] Rozas O, Vidal C, Baeza C, Jardim WF, Rossner A. Organic micropollutants (OMPs) in natural waters: Oxidation by UV/H_2O_2 treatment and toxicity assessment. Water Research. 2016;**98**(554):109-118

[12] Chen Y, Sun Z, Yang Y, Ke Q. Heterogeneous photocatalytic oxidation of polyvinyl alcohol in water. Photochemistry and Photobiology. 2011;**142**(1):85-89

[13] Kaczmarek H, Kaminska A, Swiatek M, Rabek JF. Electrochemical oxidation of polyvinyl alcohol using a RuO_2/Ti anode. Angewandte Makromolekulare Chemie. 1998;**4622**:109-121

[14] Aarthi T, Shaama M, Madras G. Degradation of water-soluble polymers under combined ultrasonic and ultraviolet radiation. Industrial and Engineering Chemistry Research. 2007;**46**:6204-6210

[15] Kim S, Kim T, Park C, Shin E. Photo-oxidative degradation of some water-soluble polymers in the presence of accelerating agents. Desalination. 2003;**155**(1):49-57

[16] Ghafoori S, Mehrvar M, Chan P. Photoreactor scale-up for degradation of polyvinyl alcohol in aqueous solution using UV/H_2O_2 process. Chemical Engineering Journal. 2014;**245**:133-142

[17] Hamad D, Mehrvar M, Dhib R. Experimental study of polyvinyl alcohol degradation in aqueous solution by UV/H$_2$O$_2$ process. Polymer Degradation and Stability. 2014;**103**:75-82

[18] McCoy B, Madras G. Degradation kinetics of polymers in solution: Dynamics of molecular weight distributions. AIChE Journal. 1997;**43**(3):802-810

[19] Tayal A, Khan S. Degradation of a water-soluble polymer: Molecular weight changes and chain scission characteristics. Macromolecules. 2000;**33**:9488-9493

[20] Madras G, Smith S, McCoy B. Degradation of poly (methyl methacrylate) in solution. Industrial & Engineering Chemistry Research. 1996;**35**(6):1795-1800

[21] McCoy B, Wang M. Continuous-mixture fragmentation kinetics: Particle size reduction and molecular cracking. Chemical Engineering Science. 1994;**49**(22):3773-3785

[22] Kodera Y, McCoy B. Distribution kinetics of radical mechanisms: Reversible polymer decomposition. AIChE Journal. 1997:3205-3214

[23] Kommineni S, Chowdhury Z, Kavanaugh M, Mishra D, Croue J. Advanced oxidation of methyl-tertiary butyl ether: Pilot study findings and full-scale implications. Journal of Water Supply: Research and Technology. 2009;**57**(1):403-418

[24] Kosaka K, Yamada H, Matsui S, Echigo S, Shishida K. Comparison among the methods for hydrogen peroxide measurements to evaluate advanced oxidation processes: Application of a spectrophotometric method using copper (II) ion and 2,9-dimethyl-1,10-phenanthroline. Environmental Science & Technology. 1998;**32**:3821-3824

[25] Ghafoori S, Mehrvar M, Chan P. Kinetic study of photodegradation of water-soluble polymers. Iranian Polymer Journal. 2012;**21**:869-876

[26] Kodera Y, Cha W, McCoy B. Continuous-kinetic analysis for polyethylene degradation. ACS Symposium Series. 1997;**42**(4):1003-1006

[27] Konaganti V, Madras G. Photocatalytic and thermal degradation of poly(methyl methacrylate), poly (butyl acrylate), and their copolymers. Industrial and Engineering Chemistry Research. 2009;**48**(4):1712-1718

[28] Shukla B, Daraboina N, Madras G. Ultrasonic degradation of poly (acrylic acid). Polymer Degradation and Stability. 2009;**94**(8):1238-1244

[29] Song H, Hyun J. An optimization study on the pyrolysis of polystyrene in a batch reactor. Journal of Chemical Engineering. 1999;**16**(3):316-324

[30] Buxton G, Greenstock C, Helman W, Ross A. Critical review of rate constants for reactions of hydrated electrons, hydrogen atoms and hydroxyl radicals ($^{\bullet}$OH/$^{\bullet}$O) in aqueous solution. Journal of Physical and Chemical Reference Data. 1988;**17**:513-886

[31] Koppenol W, Butler J, Van Leeuwen W. The Haber—Weiss cycle. Photochemistry and Photobiology. 1978;**28**:655-658

[32] Weinstein J, Bielski B. Kinetics of the interaction of HO$_2$ and O^{2-} radicals with hydrogen peroxide: The Haber-Weiss reaction. American Chemical Society. 1979;**101**:58-62

[33] Christensen H, Sehested K, Corfitzen H. Reactions of hydroxyl radicals with hydrogen peroxide at ambient and elevated temperatures. Physical Chemistry. 1982;**86**:1588-1590

[34] Bielski B, Cabelli D. Highlights of current research involving superoxide and perhydroxyl radicals in aqueous solutions. International Journal of Radiation Biology. 1991;**59**:291-319

[35] Elliot A, Buxton G. Temperature dependence of the reactions OH + O^{2-} and OH + HO_2 in water up to 200°C. Chemical Society. 1992;**88**:2465-2470

[36] Linden K, Sharpless C, Andrews S, Atasi K, Korategere V, Stefan M, et al. Innovative UV Technologies to Oxidize Organic and Organoleptic Chemicals. London: IWA Publishing; 2005

[37] Whittmann G, Horvath I, Dombi A. UV-induced decomposition of ozone and hydrogen peroxide in the aqueous phase at pH 2-7. Ozone Science and Engineering. 2002;**24**:281-291

[38] Hamad D, Mehrab M, Dhib R. Photochemical kinetic modeling of degradation of aqueous polyvinyl alcohol in a UV/H_2O_2 photoreactor. Journal of Polymers and the Environment. 2018;**26**(8):3283-3293

[39] Taghizadeh M, Yeganeh N, Rezaei M. The investigation of thermal decomposition pathway and products of poly(vinyl alcohol) by TG-FTIR. Applied Polymer Science. 2015;**32**(25): 42117-42129

[40] Liao C, Gurol M. Chemical oxidation by photolytic decomposition of hydrogen peroxide. Environmental Science & Technology. 1995;**29**:3007-3014

[41] Stefan M, Hoy A, Bolton J. Kinetics and mechanism of the degradation and mineralization of acetone in dilute aqueous solution sensitized by the UV photolysis of hydrogen peroxide. Environmental Science & Technology. 1996;**30**:2382-2390

[42] Peng Z, Kong L. A thermal degradation mechanism of polyvinyl alcohol/silica nanocomposites. Polymer Degradation and Stability. 2007;**92**: 1061-1071

[43] Bird R, Stewart W, Lightfoot E. Transport Phenomena. 2nd ed. New York: Wiley & Sons, Inc.; 1960

[44] Labas D, Zalazar S, Brandi J, Martín A, Cassano E. Scaling up of a photoreactor for formic acid degradation employing hydrogen peroxide and UV radiation. Helvetica Chimica Acta. 2002;**85**:82-95

[45] McCoy B, Madras G. Chemical Engineering Science. 2001;**56**:2831-2836

[46] Stickle J, Griggs A. Mathematical modeling of chain-end scission using continuous distribution kinetics. Chemical Engineering Science. 2012;**68**: 656-659

[47] Sterling J, McCoy B. Distribution kinetics of thermolytic macromolecular reactions. AIChE Journal. 2001;**7**:2289-2303

[48] Randolph A, Larson M. Theory of Particulate Processes: Analysis and Techniques of Continuous Crystallization. New York: Academic Press; 1971

[49] Sezgi A, Cha S, Smith J, McCoy B. Polyethylene pyrolysis: Theory and experiments for molecular weight distribution kinetics. Industrial and Engineering Chemistry Research. 1998; **37**(7):2582-2591

[50] Marimuthu A, Madras G. Continuous distribution kinetics for microwave-assisted oxidative degradation of poly(alkyl methacrylates). AIChE Journal. 2008;**54**(8):2164-2173

[51] Vinu R, Madras G. Photocatalytic degradation of polyacrylamide-coacrylic acid. Polymer Degradation and Stability. 2008;**93**(8):1440-1449

[52] Madras J, McCoy B. Oxidative degradation kinetics of polystyrene in solution. Chemical Engineering Science. 1997;**52**(16):2707-2713

[53] Smagala T, McCoy B. Mechanisms and approximations in macromolecular

reactions: Reversible initiation, chain scission, and hydrogen abstraction. Industrial and Engineering Chemistry Research. 2003;**42**:2461-2469

[54] Hulburt H, Katz S. Some problems in particle technology: A statistical mechanical formulation. Chemical Engineering Science. 1964;**19**:555-574

[55] Mehrvar M, Anderson W, Moo-Young M. Photocatalytic degradation of aqueous organic solvents in the presence of hydroxyl radical scavengers. International Journal of Photoenergy. 2001;**3**(4):187-191

Section 3

Environmental Assessment

Herbicides Mechanisms Involved in the Sorption Kinetic of Ionisable and Non Ionisable Herbicides: Impact of Physical/ Chemical Properties of Soils and Experimental Conditions

Lizethly Caceres Jensen, Angelo Neira-Albornoz and Mauricio Escudey

Abstract

Volcanic ash-derived soils (VADS, variable-charge soils) are predominant in some regions of the world, being of great importance in the agricultural economy of several emerging countries. Their amphoteric surface charge characteristics confer physical/chemical properties different to constant surface charge-soils, showing a particular behavior in relation to the herbicide adsorption kinetics. Volcanic soils represent an environmental substrate that may become polluted over time due to intensive agronomic uses. Solute transport models have contributed to a better understanding of herbicide behavior on variable and constant-charge soils, being also necessary to evaluate the fate of herbicides and to prevent potential contamination of water resources. The following chapter is divided into four sections: physical/chemical properties of variable and constant-charge soils, kinetic adsorption models frequently used to obtain kinetic parameters of herbicides on soils, solute transport models to describe herbicide adsorption on VADS, and impact of experimental conditions of kinetic batch studies on solute transport mechanisms.

Keywords: variable-charge soils, constant-charge soils, kinetic adsorption, herbicides and solute transport mechanism

1. Introduction

Nature of soils is regulated by various soil-forming factors such as parent material, climate, vegetation, and time [1]. These factors vary widely among region, and also these vary in their properties. Volcanic ash-derived soils (VADS) are predominantly found in regions of the world with geochemical characteristics dominated by active and recently extinct volcanic activity. These have great importance in the agricultural economy of several emerging and developing countries of Europe, Asia, Africa, Oceania, and America. They are abundant and widespread in central-southern Chile (from 19° to 56° S latitude), accounting for approximately 69% of the arable land [2].

Agricultural practices developed in Chilean VADS have led to the very increased use of herbicides and also require frequent adjustments of soil pH and mineral fertilization [3–5]. Among these soils, andisols and ultisols are the most abundant, both presenting an acidic pH (4.5–5.5). Andisols are rich in organic matter (OM), with high specific surface area, P retention (>85%), and variable charge with low saturation of bases, low bulk density (<0.9 Mg m^{-3}) associated with a high porosity, and a strong microaggregation of heterogeneous forms and a mineralogy dominated by short-range ordered minerals, such as allophane ($Al_2O_3SiO_2 \times nH_2O$) [2, 6]. Allophane plays a key role in surface reactivity in andisols determining the availability of nutrients and controlling soil contaminant behavior [6]. Ultisols have a low amount of OM, relatively high amounts of Fe oxides in different degrees of crystallinity, low base saturation (<30%), high bulk density (0.8–1.1 Mg m^{-3}), and high clay content (>40%) [2, 7]. This last component provides a finer texture that allows a greater cohesion with respect to andisols [2, 8].

Andisols present variable surface charge, originated in both inorganic and organic constituents. Inorganic minerals such as goethite (FeOOH), ferrihydrite ($Fe_{10}O_{15} \times 9H_2O$), gibbsite ($Al(OH)_3$), imogolite, and allophane contribute through the dissociation of Fe—OH and Al—OH active surface groups, while OM through the dissociation of its functional groups (mainly carboxylic and phenolic) and humus-Al and Fe complexes with amphoteric characteristics contributes too. For the other side, ultisols present little or no charge, because more crystalline minerals, such as halloysite and/or kaolinite dominate their mineralogy.

Several herbicide adsorption kinetic studies on VADS have indicated that the herbicide adsorption is a nonequilibrium process [5, 9]. Time-dependent adsorption can be a result from physical and chemical nonequilibrium and intrasorbent diffusion can occur during the transport of pesticides in soils [10]. In general, nonequilibrium adsorption has been attributed to several factors, such as: diffusive mass transport resistances, nonlinearity in adsorption isotherms, adsorption-desorption nonsingularity and rate-limited adsorption reactions [11]. The *intra-OM-diffusion* has been suggested to be the predominant factor responsible for the nonequilibrium adsorption of nonionic or hydrophobic compounds on VADS [9, 12]. The differences in the *intra-OM* adsorption kinetics of herbicides were due to soil constituents, such as organic carbon (OC) and mineral composition on VADS.

The adsorption-desorption behavior of pesticides is the principal process affecting the fate of these chemicals in soil and water. In general, adsorption-desorption processes are known to be important because they are time-dependent and with considerable ecosystem impact, influencing the availability of organic pollutants for plant uptake, microbial degradation, and transport in soil and consequently leaching potential. In this sense, the principal process that affects the fate of pesticides in soil and water is adsorption of pesticides from soil solution to soil particle active sites, which limit transport in soils by reducing their concentration in the soil solution. Therefore, adsorption kinetic studies provide important information for weed control, crop toxicity, runoff, and carryover events, serving as the foundation for estimating effects on biotic and abiotic environmental components. The kinetic parameters can be obtained by means of the application of two kinds of kinetic models: the ones that allow establishing principal kinetic parameters and modeling of the adsorption process and other models frequently used to describe adsorption mechanisms of organic compounds on soils. Such information is necessary in order to understand leaching of herbicides for preventing potential contamination of groundwater.

The aim of this chapter is to establish the differences of adsorption kinetics of ionizable and nonionizable herbicides (INIH) in Chilean VADS to investigate the mechanisms involved of INIH adsorption on VADS by applying different solute adsorption mechanism models. Kinetic adsorption model description is also necessary in order to develop and validate computer simulation transport models on VADS to prevent potential contamination of water resources, considering model restrictions related to experimental conditions of kinetic batch studies on solute transport mechanisms.

2. Physical and chemical properties of variable-charge soils

Variable-charge soils are dominated by Al/Fe-humus complexes, by ferrihydrite, a short-range-order Fe hydroxide mineral, or by clay components characterized by the formation of short-range-order aluminosilicates, such as allophane and imogolite [13]. The clay fraction mineralogy of VADS is usually dominated by allophane with a minor content of kaolinite, gibbsite, goethite, and hematite [14]. Besides, these minerals contain 2:1 and 2:1:1 type minerals and their integrades, opaline silica and halloysite [13].

These distinctive physical and chemical properties are largely due to the formation of noncrystalline materials, biological activity, and the accumulation of OM [13, 15]. The soil OM represents a key indicator of soil quality, both for agricultural (i.e., productivity and economic returns) and environmental functions (i.e., carbon sequestration). Andisols are highly representative of VADS; their OC concentration is more associated with metal-humus complexes than with concentrations of noncrystalline materials. Nevertheless these materials with variable charge surfaces provide an abundance of microaggregates that permit to encapsulate OC, favoring their physical protection [13]. Other studies indicate that Al/Fe oxides/hydroxides in allophanic soils are linked to carboxylic and aromatic groups of soil OM being the last highly decomposed [1].

In general, andisols are soils rich in constituents with amphoteric surface reactive group being considered the most abundant variable charge soils in Chile [14]. The most striking and unique properties of these are: variable charge, high water-holding capacity, low bulk density, high friability, highly stable soil aggregates due to unstable colloidal dispersions, excellent tilth and strong resistance to water erosion [13], anion adsorption, high lime or gypsum requirement to achieve neutral pH, and considerable adsorption affinity for cations (Ca and Mg), which may form both inner- and outer-sphere complexes although the first are found to be more important [14].

Andisols are relatively young soils and cover about 0.84% of the world's land [13, 16], being a typical product of weathering increases in temperate and tropical environments with sufficient moisture [13]. In this sense, metastable noncrystalline materials are transformed to more stable crystalline minerals (e.g., halloysite, kaolinite, and gibbsite) allowing the alteration of andisols to Inceptisols, alfisols, or ultisols. Andisols are often divided into two groups based on the mineralogical composition of A horizons: allophanic andisols dominated by variable charge constituents (allophane/imogolite), and nonallophanic andisols dominated by both variable charge and constant charge components (Al/Fe-humus complexes and 2:1 layer silicates) [13]. Allophanic andisols form preferentially in weathering environments with pH values in the range of 5–7 and a low content of complexing organic compounds. Nonallophanic andisols form preferentially in pedogenic environments that are rich in OM and have pH values of 5 or less [13].

Allophanic andisols present allophane, imogolite, poorly crystalline Fe oxides (probably ferrihydrite), Al/Fe-humus complexes, volcanic glass (which is a mixture of aluminosilicates and traces of ferromagnesian minerals), and secondary Si minerals (opaline silica), resulting in pH-dependent variable charge, CEC and anion exchange capacity (AEC), and high phosphate retention >70% [1, 13]. Allophane, the main component of the clay fraction of VADS, has short to mid-range atomic order and a prevalence of Si—O—Al bonding [17]. This aluminosilicate consists of hollow, irregularly spherical nanoparticles with an outside diameter of 3.5–5.0 nm, a wall thickness of 0.7–1 nm, and a specific surface area of 700–900 $m^2\,g^{-1}$ with a chemical composition generally ranging from an Al:Si atomic ratio of 1:1–2:1 [13].

The presence of allophane in andisols provides excellent physical fertility properties for crop production, such as: high friability, stable aggregates, ease of root penetration, good drainage, high permeability, low bulk density at field-moisture water content <0.9 $g\,cm^{-3}$, high porosity, and high air and water retention [13]. An unusually high amount of micropores in allophanic VADS is partially attributable to the intra- and inter-particle pores of allophane [15]. The development of aggregates in VADS is closely related to the retention of large amounts of plant-available water. The large volume of both mesopores/micropores relates with the high water-holding capacity of andisols. In this sense, young VADS have a greater amount of macropores larger than 100 μm in diameter and a lesser amount of mesopores (0.4–6.0 μm) and micropores (<0.4 μm). In contrast, moderately weathered soils have a large amount of mesopores (0.4–6.0 μm) and micropores (<0.4 μm), contributing to the large plant-available water.

Based on their surface charge characteristics, VADS are characterized by a mixed charge system [14]. In this sense, the soil particles are of two different types: dual and variable-charge particles (phyllosilicates and allophane) and variable-charge particles (Fe/Al oxides). The surface charge density of variable-charge oxides depends on pH and ionic strength (IS) of the soil solution. The Fe/Al oxides have a surface reactive group with amphoteric properties; these groups are protonated and positively charged under acidic conditions (at a pH below the point of zero charge, PZC) or deprotonated and negatively charged under basic conditions (at a pH higher than the PZC). In general, the PZC of Al/Fe oxides are between 8 and 9. The Fe in VADS is present mostly in the form of noncrystalline hydroxides (ferrihydrite) and partly as Fe-humus complexes [13]. Ferrihydrite appears as individual spherical particles ranging in size between 2 and 5 nm. These particles form aggregates ranging from 100 to 300 nm in diameter [13].

Dual-charge particles, such as phyllosilicates and allophane, usually develop permanent and variable charge or only variable charge but with different magnitude an even different sign on different surfaces of the same particle. These inorganic minerals are abundant in VADS, controlling chemical properties of the bulk soil. The siloxane ditrigonal cavity of the phyllosilicate siloxane surfaces may develop a localized permanent negative charge as a result of isomorphic substitutions in their internal crystal structures regardless of ambient conditions. The magnitude of this permanent negative charge does not depend on pH and IS of the soil solution. In contrast, the edges of these particles develop variable charge.

The variable charge of allophane is the result of protonation and dissociation of Al—OH and Si—OH superficial functional groups, with Al—OH groups having negative, neutral, or positive charge and the more acidic Sl—OH groups having either neutral or negative charge. As allophane is a dual-variable charge, VADS usually have a slightly acidic to acidic soil solution pH [14]. Under acidic conditions, the surfaces of these minerals are net positively charged. The variable positive charge results from protonation of surface inorganic soil constituents with Al—OH, Fe—O, and Fe—OH groups, while the variable negative charge results from dissociation of surface Si—OH and organic functional groups of organic soil constituents (e.g., carboxylic, phenolic, or amino reactive groups) [13]. The development of negative charge with increasing soil pH has been common to all andisols and has been strongly related to the amount of soil OM [13]. Soils with a large variable charge component required large additions of lime for pH amendment and were susceptible to leaching of cations when the soil pH decreased [13]. The CEC and AEC of variable charge components on particle edges are pH- and IS-dependent of the soil solution. On the other hand, the most important mineralogical components in ultisols are: kaolinite (dual-charge minerals), hydroxy-interlayered vermiculite, muscovite, smectite, and Fe/Al oxides (quartz in the sand and silt fractions). Kaolinite is a 1:1 phyllosilicate with a relatively low specific surface area (between 5 and 39 $m^2 g^{-1}$) and presents the lowest surface charge (about 1–5 cmol (c) kg^{-1}) among common dual-charge clay minerals. The PZC of kaolinite is between 2.8 and 2.9 [14]. The kaolinite in A horizons has the same tubular morphology as the halloysite at depth suggesting that hydrated halloysite transforms to kaolinite upon dehydration. Halloysite is a 1:1 aluminosilicate hydrated mineral characterized by a diversity of morphologies (e.g., spheroidal and tubular), specific surface area, structural disorder, and physical-chemical properties (e.g., cation exchange capacity (CEC) and ion selectivity) [13].

3. Physical and chemical properties of constant-charge soils

There are different soil orders classified by soil formation, climates, and morphological features [18]. However, globally, most of the soil orders have constant charge. In general, these soils present a similar composition to the andisols, except for amorphous clays, metal oxides, oxyhydroxides, and hydroxides. In this sense, constant-charge soils are a simplification of andisols, what is expressed in a lower variety of adsorbent forms that result in a minor mechanistic variability of adsorption-desorption processes. The lack of Fe/Al oxides and allophane involves a surface without humus-Fe/Al, Fe/Al-mineral, and mineral-Fe/Al-humus complexes, reducing the combinations of possible surface-surface and herbicide-surface interactions, increasing the colloidal stability due to electrostatic repulsion between non-Fe/Al minerals and OM, both with surfaces dominated by anionic sites (S^-).

A thorough analysis is required to study the adsorption kinetics with agricultural or remediation purposes. For example, histosols from peat or bog have a high OM content (>20%) [18], so the adsorption process can be simplified to the soil/solution partition coefficient normalized to the OC content (K_{oc}) or OM content (K_{om}) [19], and the stability of microaggregates by OM. Aridisols, developed in arid regions, have a high presence of clay and salts such as sodium, calcium carbonates, or gypsum, together with a low water content [18], so the adsorption process can be simplified to clay/solution partition coefficient, with high probability of equilibrium and/or precipitation of adsorbate under field conditions. Ultisols, with low base saturation but high clay content, OM and acidity, have humus as the main soil component that contributes to the little variable charge on these soils, controlling the pesticide adsorption mainly through hydrophobic and H-bonding.

Despite the diversity previously exposed, in all the cases, the adsorption sites are mostly neutral (S^0, e.g., $OM_{aromatic,aliphatic}$) and anionic (S^-, e.g., siloxane), and this implies adsorption of hydrophobic, polar, and cationic herbicides, where the dominance of siloxane and anionic organic surface groups generates a low PZC and negative surface charge, mostly pH-independent [20], which therefore implies small changes in CEC of minerals and negligible AEC at soil pH. So, the adsorption is independent of PZC for constant-charge soils. We will use ultisols as a constant-charge soil to show this and contrast with andisols.

3.1 Effect of MSM adsorption in PZC on ultisols and andisols

The curve of PZC versus pH for ultisol and andisol soils is shown in **Figure 1** [5]. As can be observed, a displacement of PZC to a higher pH was produced in both soil surfaces with adsorbed metsulfuron-methyl (MSM) confirming the contribution of

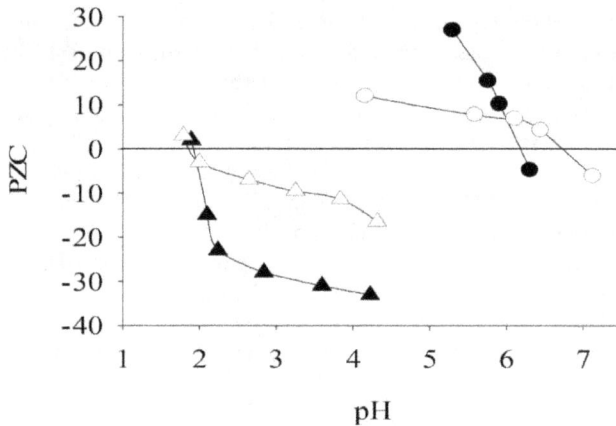

Figure 1.
Electrophoretic migration curves: (▲) ultisol without MSM adsorbed; (△) ultisol with 15 µg mL^{-1} of MSM adsorbed; (●) andisol without MSM adsorbed; and (○) andisol with 15 µg mL^{-1} of MSM adsorbed [5].

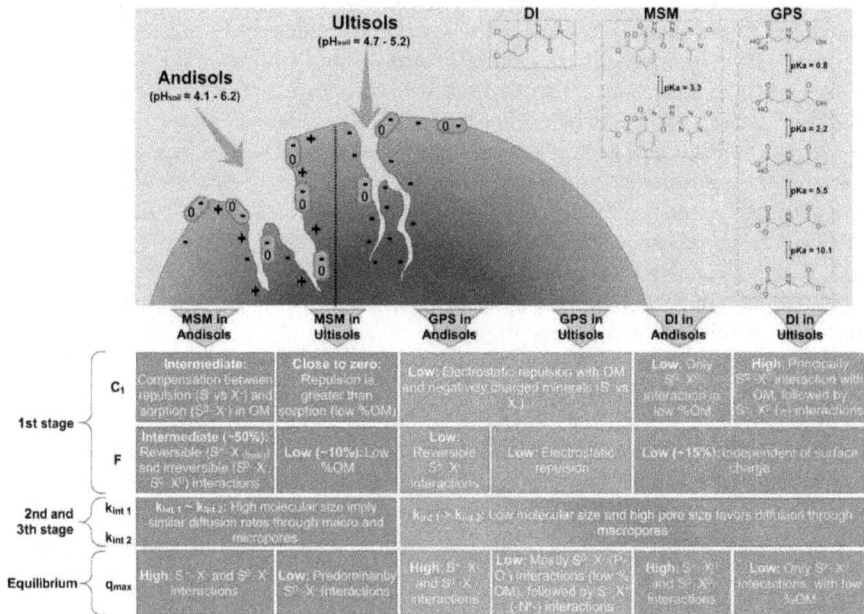

Figure 2.
*Connection between adsorption parameters and mechanistic explanation from kinetic models in andisols and ultisols from **Figure 3** (S$^+$, S$^-$, and S^0: surface charge. X$^+$, X$^-$, and X^0: herbicide species. The green ovals are OM. pH$_{soils}$ were measured in water at 1:2.5 soil:solution ratio).*

charged surface sites to adsorption of anionic MSM through electrostatic and hydrophilic interactions on ultisols and andisols, respectively. The OM and active and free Fe/Al oxides will control the adsorption process in andisols mainly through hydrophilic on surface minerals, such as allophane, gibbsite, hematite, and goethite. In contrast, andisols present positive sites (S$^+$, e.g., goethite at pH < 7.8) in addition to S^0 and S$^-$, that allow the anionic herbicide adsorption (X$^-$(hydr) and X$^-$) (**Figure 2**). Some intuitive mechanisms affected by pH, pK$_a$, and PZC are anionic and cationic exchange due to their electrostatic nature, but these kinds of adsorption are usually accompanied by other mechanisms.

4. Adsorption kinetics

The adsorption is characterized by a three-stage process: a rapid uptake on readily available adsorption sites (**Figure 2**, 1st stage), followed by slow diffusion-immobilization into mesopores, micropores, or capillaries in the sorbent's internal structure through mechanisms controlled by *intraparticle diffusion* (*IPD*), with small adsorbed amounts appearing on the external surface [9] (**Figure 2**, 2nd stage). The third stage (**Figure 2**), which is controlled by mass transfer mechanisms, involves rapid uptake of the solute in the inner surface of the sorbent. Adsorption processes are known to be important because they are time-dependent with considerable ecosystem impact, influencing the availability of organic pollutants for plant uptake, microbial degradation, and transport of pesticides in the soil environment during the short-term and consequently affecting their leaching potential [4, 9, 12, 21]. Adsorption kinetic studies may provide important information related to weed control, crop toxicity, runoff, and solute transport mechanisms [9, 22].

4.1 Kinetic adsorption models frequently used to estimate kinetic parameters of herbicides on soils

The *pseudo-second order* (*PSO*) model (**Figure 3A**) has been the best adsorption kinetic model to establish principal kinetic parameters and modeling of the adsorption process of INIH on ultisols and andisols [9, 12, 23]. In the **Figure 3A**, q_t is the adsorbed quantity ($\mu g\ g^{-1}$) at any soil-solution contact time t (min) for kinetic adsorption experiments. The higher value of the overall rate constant k_2 of MSM with respect to DI adsorption on andisols indicates that this value reflects contributions from the favored electrostatic interactions considering both a retarded *IPD* as well as *intra-OM-diffusion*.

4.2 Mechanistic kinetic models to describe herbicide adsorption on ultisols and andisols

Figure 3B and **C** show a different transport mechanism for glyphosate (GPS, pK_a = 0.8; 2.23; 5.46; 10.14), metsulfuron-methyl (MSM, pK_a = 3.3) and diuron (DI) on ultisol and andisols [9, 12, 23]. The *IPD* or Weber-Morris model (**Figure 3B**) is one of the most used models to describe solute transport mechanisms of organic compounds in different adsorbents intended for remediation purposes [9, 12, 23]. Nevertheless, the *two-site nonequilibrium* (*TSNE*) model (**Figure 3C**) has been the

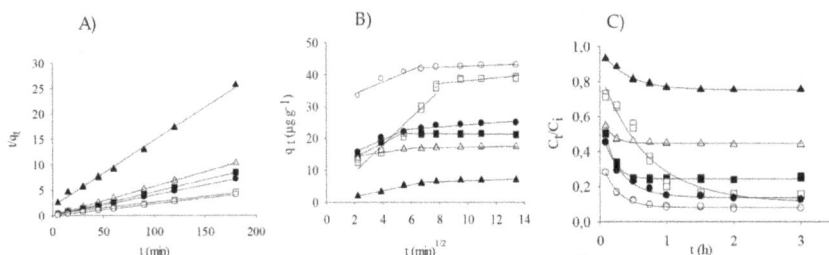

Figure 3.
(A) PSO model for adsorption kinetics of GPS on ultisol (■) and andisol (□), MSM on ultisol (▲) and andisol (△) and DI on ultisol (●) and andisol (○). (B) IPD model for GPS adsorption kinetics on ultisol (■) and andisol (□); MSM on ultisol (▲) and andisol (△); and DI on ultisol (●) and andisol (○). (C) TSNE model plots for GPS adsorption on ultisol (■) and andisol (□), MSM adsorption on ultisol (▲) and andisol (△) and DI adsorption on ultisol (●) and andisol (○) [9, 12]. $pH_{ultisol}$ = 5.2 and $pH_{andisol}$ = 4.1.

best kinetic model to describe INIH transport mechanisms on VADS [9, 12]. In **Figure 3C**, c_t is the solute concentration at any time ($\mu g\ mL^{-1}$) and c_{in} is the initial added solute concentration ($\mu g\ mL^{-1}$). As an advantage, the *TSNE* model allows obtaining the soil/solution partition coefficient (K_d), percentage of instantaneous adsorption (*F*), and first-order desorption rate constant (k_{des}) from the time-dependent adsorption sites ($100 - F$).

Time-dependent adsorption can be a result of physical and chemical nonequilibrium [12]. The nonequilibrium adsorption on soils has been attributed to several factors, such as: diffusive mass transfer resistances, nonlinearity in adsorption isotherms, adsorption-desorption nonsingularity, and rate-limited adsorption reactions [11]. The rate-limited diffusion of the adsorbate from bulk solution to the external surface of the sorbent and rate-limited diffusion within mesopores and micropores of the soil matrix will occur before the equilibrium is reached. The adsorption process first occurs within the boundary layer around the sorbent being conceptualized as a rapid uptake process on readily available adsorption sites. The intercept of the first adsorption step (c_1) by the *IPD* model has been related to the extent of the boundary layer effect, namely, the diffusion of solute molecules through the solution to the external surface of the adsorbent [21]. In this regard, c_1 is the initial solute adsorption (mg/g) [24] being proportional to the thickness of the boundary layer.

When $c_1 = 0$, the rate of adsorption is controlled by only *IPD* for the entire adsorption period (**Figure 2**, 1st stage and **Figure 3B**) [9, 12, 21]. In this regard, an external mass transfer of the solute from the bulk solution to the soil particle surface exists. This can be seen in the initially steeper linear ($kint_1$) (**Figure 2**, 1st stage), where the MSM adsorption on ultisols was controlled exclusively by *IPD* (c_1 close to 0) (**Figure 3B**). A low value for c_1 has been related with the heterogeneity of the adsorbent, enhanced by the colloidal stability of ultisols, implying a high meso- and micro-porosity with a complex pore morphology. In this regard, the intercept also decreases with the increasing surface heterogeneity of the soils, indicating a small film resistance to mass transfer surrounding the adsorbent particle. This can be seen in the initially steeper linear ($kint_1$) (**Figure 2**, 2nd stage), where the mass transfer across the boundary layer and *IPD* were the two mechanisms to control the MSM adsorption on andisols. The last mechanisms were observed for GPS and DI adsorption on ultisols and andisols (**Figure 3B**), where GPS presented the same initial adsorption on both kinds of soils. A large value for c_1 indicates that the adsorption proceeds via a more complex mechanism consisting of both surface adsorption and *IPD*. In this regard, the highest c_1 values for DI correspond to large film diffusion resistance due to the greater boundary layer effect surrounding the particles for DI [24, 25] indicating a rapid adsorption in a short time and highest initial DI adsorption on andisols with a wide distribution of pore sizes. This was associated with the macroporosity of andisols due to aggregates composed by humus-Al/Fe complexes with low colloidal stability. If andisols have high OM content, these complexes could generate a preferential flow, increasing the transport of herbicides by the remaining colloidal or dissolved OM, increasing the leaching potential of DI.

The second ($kint_1$) and ($kint_2$) third stage describe the gradual adsorption stage (**Figure 2**, 2nd and 3rd stage), where *IPD* through macropores is rate limiting [21] followed by slow diffusion-immobilization in micropores or capillaries of the sorbent's internal structure. The rate-controlling step may be controlled by *film diffusion* and *IPD* [26]. In this sense, the adsorption process proceeds in the liquid-filled pore (external mass transfer (EMT)) steps or along the walls of the pores of the sorbent (internal mass transfer (IMT)) steps. It is assumed that the external resistance to mass transfer surrounding the solute is significant only in the early stage of adsorption [27]. This can be seen in the initially steeper linear ($kint_1$) (**Figure 2**, 2nd stage), where the *IPD* model indicated that mass transfer across the boundary layer and *IPD* to control the GPS and DI adsorption on ultisols and andisols (**Figure 3A**). The molecules of GPS diffuse quickly through the macropores of the andisols. MSM adsorption on ultisols was controlled exclusively by macropore *IPD* (**Figure 3A**) and MSM adsorption on andisols occurred on two stages ($kint_1 \approx kint_2$) being controlled by *EMT* and *IPD*.

The 3rd stage (**Figure 2**) is the adsorption of the particle in the inner surface of the sorbent through mass-action-controlled mechanisms where a rapid uptake occurs or mechanism of surface reaction which consider the interactions between functional groups of solute and surface (as a chemisorption) [9, 12, 24]. In this regard, this stage is observed in the second linear portion ($kint_2$, **Figure 2**, 2nd stage and **Figure 3B**) being the gradual adsorption stage in which *IPD* dominates [27]. The second line (**Figure 2**, 3rd stage and **Figure 3B**) will depict micropore diffusion with the IMT occurring during the retention of INIH. In general, INIH present a highest adsorption capacity (q_{max}) on andisols. In this regard, the OM contributes to the ionic adsorption through the dissociation of its functional groups (mainly carboxylic and phenolic) and Al/Fe-humus complexes with amphoteric characteristics. And also, the organic colloids have an important role in hydrophobic pesticide transport [28], due to their small size (<0.45 μm) and high affinity to nonpolar functional groups. This could have implications for the management of soils and pesticides in relation to the release of organic colloids into solution, especially in rainy areas or organic soils. The exogenous and endogenous water-extractable OM (WEOM) can influence the pesticide transport on soils, through formation of WEOM-pesticide complexes or competition between WEOM and pesticides for the adsorption sites, and thus, the retention and transport of the pesticides decrease and increase, respectively [29]. In this regard, Thevenot et al. [29] found that on a sandy-loam soil with low DI and WEOM adsorption capacity, the application of organic amendments with high WEOM content could increase DI leaching and, consequently, ground-water contamination risks.

While inorganic minerals such as goethite, ferrihydrite, gibbsite, imogolite, and allophane contribute through the dissociation of Si—OH, Fe—OH, and Al—OH active surface groups [21], kaolin clays could contribute to the adsorption in ultisols through Si—OH and (Al—OH—Si)$^{+0.5}$ from the exposed edge kaolinite of the octahedral and tetrahedral basal surfaces having a hydrophilic and hydrophobic character, respectively [21].

For the case of ionizable herbicides, such as MSM, a negative correlation has been found between adsorption capacity and pH on acidic andisols (pH$_{soils}$ between 4.49–6.46) from southern China [30] and acidic ultisols (pH$_{soils}$ acidic 4.7–5.2) and acidic andisols (pH$_{soils}$ acidic 4.1–6.2) from Chile [5, 31]. This behavior could be related to the adsorption mechanism of neutral pesticide species (X^0) on OM at low pH ($S^{0...}X^0$, with S^0 = hydrophobic or polar OM) and the increase of repulsion between S^- and X^- at high pH (**Figures 1** and **2**, 1st stage). In addition, a positive correlation between adsorption capacity and CEC and amorphous and free Al/Fe content even at low pH implies a significant electrostatic adsorption mechanism on S^+, probably anionic exchange ($S^{+...}X^-(hydr)$) (**Figures 1** and **2**). The presence of two mechanisms explains the variations on K_{oc} value, and the reversibility of anionic exchange explains the lower hysteresis at higher pH.

For the case of amphoteric herbicides, such as imazaquin, Weber et al. studied the imazaquin adsorption (pK$_a$ = 1.8 for —NH$^+$—, 3.8 for —COO$^-$, and 10.5 for —N—) in Cape Fear soil [32]. In this acidic ultisol (pH$_{soil}$ = 4.7), the authors found a positive correlation between adsorption capacity and presence of cationic (X^+) and neutral imazaquin, attributed to cationic exchange, an electrostatic mechanism opposite to anionic exchange observed for MSM in andisols ($S^{-...}X^+(hydr)$)

instead of $(S^{+\cdots*}X^-(hydr))$, while the inverse correlation with the anionic species was explained by electrostatic repulsion with negative charge surfaces. In this sense, the different surface charge of both soil orders (predominance of S^- for ultisols and S^+ for andisols) plays an important role in herbicide-soil speciation that should be considered together with molecular properties to explain the mechanistic behavior of adsorption.

On the other hand, the OM, humic substances, and clay content increased the adsorption and reduced the mobility of imazaquin at low pH, due to the inverse relationship between adsorption and transport [32]. But this trend involves the adsorption on negatively charged sites, then andisols could exhibit a different behavior affected by positive charges and their interactions with herbicides $(S^{+\cdots*}X)$ and OM $(S^{+\cdots*}S^-)$. For the case of a nonionizable herbicide, such as metolachlor, the OM plays a fundamental role for specific and nonspecific adsorption mechanisms. In this regard, Weber et al. studied the adsorption of metolachlor in Cape Fear soil [32], comparatively to imazaquin. In general, metolachlor was adsorbed by physical binding with soil. The proposed mechanisms were hydrophobic bonding to lipophilic sites of OM and humic substances $(X^{0\cdots*}S^0)$, charge-transfer mechanisms, van der Waals forces, and H-bonds on polar surfaces of clay minerals, with a greater adsorption than imazaquin, similar to DI in Chilean soils (**Figure 2**, 1st stage and equilibrium) [9]. Additionally, the adsorption process could be dependent on mass transfer instead of soil-herbicide affinity. In this sense, the adsorption mechanisms depend on chemical and physical properties of soil and herbicide, including the interaction between soil components. This was observed for atrazine adsorbed on OM [33], in which hydrophobic interactions were dominant in aliphatic C of the inner sites of humic self-associated aggregates for ultisols, while for andisols the adsorption occurred on the surface of aromatic C stabilized by allophane and therefore becomes more easily desorbed [33]. This effect on conformational rigidity of organic and Fe/Al-humus sorbents is interesting to predict the environmental fate of organic nonionizable herbicides, where the formation of stable Fe/Al-humus complexes becomes OM less heterogeneous in andisols, which plays an important role in controlling the reversibility of adsorption processes. The effect of soil-solution interaction on hydrophobic adsorption of acetamiprid ($pK_a = 4.16$) was studied by Murano et al. [34]. The adsorption of neutral acetamiprid at pH 6.5 (neutral specie) increased when Al^{+3} or Fe^{+3} was added to humic substances because of hydrophobicity enhanced by cation bridging in the formation of humic substance-metal complexes ($S^{-\cdots*}M^{+3\cdots*}S^-$ and $3S^{-\cdots*}M^{+3}$, where $M=Al^{+3}/Fe^{+3}$), changing the surface charge, conformational structure of humic substances, and accessibility to reactive sites. The *TSNE model* (**Figure 3C**) indicated that MSM adsorption on andisols presented an initial phase with a fast trend to equilibrium, where ~50% (F ~ 50%) of sites account for almost instantaneous equilibrium, while for ultisols, great part of sites corresponded to the time-dependent stage of adsorption (91%, F ~ 10%) (**Figure 2**, 1st stage). As F is related to irreversible adsorption, this parameter acts as an indirect indicator of hysteresis. So, high F values imply low desorption. The instantaneous adsorption on andisols was associated to OM-pesticide complexes, more stable and irreversible than clay-pesticide complexes, which was consistent with low k_{des} values. On the other hand, F on ultisols was related to a pore deformation mechanism due to the hysteresis water sorptivity in hydrated minerals, such as halloysite.

The DI adsorption on andisols presented an initial phase with a fast trend to equilibrium, where between 10 and 50% of sites account for very fast adsorption (**Figure 2**, 1st stage). Again for ultisols, most of the sites corresponded to the time-dependent stage of adsorption (90%) (**Figure 2**, 1st stage). The adsorption of nonionic or hydrophobic compounds on VADS has been described as a two-site equilibrium-kinetic process, where *intra-OM-diffusion* has been suggested to be the predominant factor responsible for the nonequilibrium adsorption [9]. In this sense, the OM is the governing factor for NIH adsorption on andisols (OC content higher than 4.0%). The presence of crystalline minerals, such as kaolinite, halloysite, and Al/Fe oxides will be significant in the *IPD* mechanism in ultisols [5, 35]. The way minerals, present on VADS, are interrelated or chemically spatially distributed, either being freely distributed throughout the soil mass or coating silt and clay grains, is determinant on their chemical role in the whole ion adsorption-desorption mechanisms [7]. The different mineral composition of VADS will have an impact on their different physical behavior, influencing the INIH adsorption rate, the adsorption mechanism involved, and the INIH adsorption capacity. All of the above must be taken into account to evaluate the potential leaching of INIH in VADS.

4.3 Impact of experimental conditions of kinetic batch studies on solute transport mechanisms

The adsorption mechanism is strongly related to the experimental conditions established to carry out the adsorption kinetic study. Considering previous examples of herbicide adsorption kinetics on ultisols and andisols exposed in **Figure 3**, the pH can affect the speciation of MSM and GPS [4, 12]. Similarly, pH can significantly affect the fraction of different soil sites for adsorption (S^+, S^- and S^0) in andisols. In this sense, the variability in q_{max} due to changes in adsorption mechanisms for the pH effect will be MSM and GPS in andisols > MSM and GPS in ultisols > DI in andisols > DI in ultisols. In addition, the cations and anions in solution, related with IS and ionic composition, can affect the CEC and AEC of soil, including the ionic exchange mechanism for MSM and GPS by (i) competition, such as GPS versus phosphate, with affinity for the same S^+, or (ii) cooperativity, such as GPS and Ca^{2+} by cation bridge adsorption in S^- ($S^{-\cdots *}Ca^{2+\cdots *}GPS^-$). To evaluate (i), initial status of soil must be known, e.g., the natural or anthropogenic P content in agricultural soils. The same driving factors mentioned above could modify the structure and porosity of soils, changing the transport mechanism of pesticides. Interactions such as $S^{+\cdots *}S^-$ in Al/Fe-humus complexes or $S^{-\cdots *}M^{+\cdots *}S^-$ and $S^{+\cdots *}X^{-\cdots *}S^+$ for the solution composition can explain the difference between $kint_1$ and $kint_2$, especially for GPS adsorption in andisols, due to the joint effect between a high content of macropores and a low molecular size. In all previous cases, different stoichiometric coefficients may be related to the same mechanism, depending on herbicide and soil properties. If we describe the adsorption rate based on the adsorption capacity only, a simple case will be the anionic exchange of GPS^- and GPS^{2-} (e.g., phosphonate group) on S^+ and $2S^+$, respectively, where $GPS^- + S^+ \rightarrow GPS-S$ is a pseudo-first order (*PFO*) reaction with respect to soil ($v \propto [S^+]$) and GPS ($v \propto [GPS^-]$), while $GPS^{2-} + 2S^+ \rightarrow GPS-S_2$ is a *PSO* reaction with respect to soil ($v \propto [S^+]^2$) but *PFO* reaction with respect to GPS ($v \propto [GPS^-]$). In this sense, it will be important to consider the experimental conditions. For example, for high soil:solution ratios, both cases will be represented by a *PFO* reaction (variation on [GPS$^-$] or [GPS^{2-}]), while an excess of herbicide or low soil:solution ratios will be represented by a *PFO* (variation on [S^{2+}]) or *PSO* reaction (variation on [S^+]).

5. Conclusions

The surface charge amphoteric characteristics of VADS confer them physical/chemical properties absolutely different to constant charge-soils, where soil composition (i.e., SOM), mineralogy, and variable charge are key components of most VADS controlling soil INIH adsorption, representing an environmental substrate that may become polluted over time due to intensive agronomic uses. The *PSO* and *TSNE* models have been the best to describe kinetics and solute transport mechanisms of INIH on VADS. These models are also necessary in order to develop and validate *QSAR* models to predict INIH adsorption on VADS to prevent potential contamination of water resources and predict environmental risks. The complex adsorption mechanisms of INIH on VADS and the diversity of soil mineralogy, texture, OC structure, and content make it necessary to consider them in *QSAR* model applications, not only to predict INIH adsorption but also to contribute to a better understanding behavior of INIH on VADS.

Acknowledgements

This work was funded via projects FONDECYT 11110421 from CONICYT, Chile, CEDENNA FB0807 (Basal Funding for Scientific and Technological Centers) from CONICYT, Chile, and PFCHA/DOCTORADO NACIONAL/2017—21170499 from CONICYT, Chile.

Conflict of interest

The authors certify that they have no conflict of interest with the subject matter discussed in this chapter.

Author details

Lizethly Caceres Jensen[1*], Angelo Neira-Albornoz[1,2] and Mauricio Escudey[3,4]

1 Laboratory of Physical and Analytical Chemistry, Department of Chemistry, Universidad Metropolitana de Ciencias de la Educación, Santiago, Chile

2 Facultad de Ciencias Químicas y Farmacéuticas, Universidad de Chile, Independencia, Chile

3 Facultad de Química y Biología, Universidad de Santiago de Chile, Santiago, Chile

4 Center for the Development of Nanoscience and Nanotechnology, CEDENNA, Santiago, Chile

*Address all correspondence to: lyzethly.caceres@umce.cl

IntechOpen

References

[1] Sarmah AK, Muller K, Ahmad R. Fate and behaviour of pesticides in the agroecosystem—A review with a New Zealand perspective. Australian Journal of Soil Research. 2004;**42**:125-154. DOI: 10.1071/sr03100

[2] Escudey M, Galindo G, Forster JE, Briceño M, Diaz P, Chang A. Chemical forms of phosphorus of volcanic ash-derived soils in chile. Communications in Soil Science and Plant Analysis. 2001;**32**:601-616. DOI: 10.1081/CSS-100103895

[3] Báez ME, Espinoza J, Silva R, Fuentes E. Sorption-desorption behavior of pesticides and their degradation products in volcanic and nonvolcanic soils: Interpretation of interactions through two-way principal component analysis. Environmental Science and Pollution Research. 2015;**22**:8576-8585. DOI: 10.1007/s11356-014-4036-8

[4] Cáceres-Jensen L, Gan J, Báez M, Fuentes R, Escudey M. Adsorption of glyphosate on variable-charge, volcanic ash-derived soils. Journal of Environmental Quality. 2009;**38**: 1449-1457. DOI: 10.2134/jeq2008.0146

[5] Caceres L, Fuentes R, Escudey M, Fuentes E, Baez MaE. Metsulfuron-methyl sorption/desorption behavior on volcanic ash-derived soils. Effect of phosphate and pH. Journal of Agricultural and Food Chemistry. 2010;**58**:6864-6869. DOI: 10.1021/jf904191z

[6] Briceno G, Demanet R, de la Luz Mora M, Palma G. Effect of liquid cow manure on andisol properties and atrazine adsorption. Journal of Environmental Quality. 2008;**37**: 1519-1526. DOI: 10.2134/jeq2007.0323

[7] Pizarro C, Fabris J, Stucki J, Garg V, Morales C, Aravena S, et al. Distribution of Fe-bearing compounds in an Ultisol as determined with selective chemical dissolution and Mössbauer spectroscopy. Hyperfine Interactions. 2007;**175**:95-101. DOI: 10.1007/s10751-008-9594-z

[8] Seguel S O, Orellana S I. Relación entre las propiedades mecánicas de suelos y los procesos de génesis e intensidad de uso. Agro Sur. 2008;**36**: 82-92. DOI: 10.4206/agrosur.2008.v36n2-04

[9] Caceres-Jensen L, Rodriguez-Becerra J, Parra-Rivero J, Escudey M, Barrientos L, Castro-Castillo V. Sorption kinetics of diuron on volcanic ash derived soils. Journal of Hazardous Materials. 2013;**261**:602-613. DOI: 10.1016/j.jhazmat.2013.07.073

[10] Brusseau ML, Rao PSC. The influence of sorbate-organic matter interactions on sorption nonequilibrium. Chemosphere. 1989;**18**:1691-1706. DOI: 10.1016/0045-6535(89)90453-0

[11] Villaverde J, van Beinum W, Beulke S, Brown CD. The kinetics of sorption by retarded diffusion into soil aggregate pores. Environmental Science & Technology. 2009;**43**:8227-8232. DOI: 10.1021/es9015052

[12] Cáceres-Jensen L, Escudey M, Fuentes E, Báez ME. Modeling the sorption kinetic of metsulfuron-methyl on Andisols and Ultisols volcanic ash-derived soils: Kinetics parameters and solute transport mechanisms. Journal of Hazardous Materials. 2010;**179**:795-803. DOI: 10.1016/j.jhazmat.2010.03.074

[13] Dahlgren RA, Saigusa M, Ugolini FC, Donald LS. The nature, properties and management of volcanic soils. In: Advances in Agronomy. Academic Press; 2004. pp. 113-182. DOI: 10.1016/S0065-2113(03)82003-5

[14] Qafoku NP, Ranst EV, Noble A, Baert G. Variable charge soils: Their mineralogy, chemistry and

management. In: Advances in Agronomy. Oxford: Academic Press; 2004. pp. 159-215. DOI: 10.1016/S0065-2113(04)84004-5

[15] Shoji S, Takahashi T. Environmental and agricultural significance of volcanic ash soils. Global Environmental Research-English Edition. 2002;**6**:113-135

[16] Takahashi T, Shoji S. Distribution and classification of volcanic ash soils. Global Environmental Research-English Edition. 2002;**6**:83-98

[17] Cea M, Seaman JC, Jara AA, Fuentes B, Mora ML, Diez MC. Adsorption behavior of 2,4-dichlorophenol and pentachlorophenol in an allophanic soil. Chemosphere. 2007;**67**:1354-1360. DOI: 10.1016/j.chemosphere.2006.10.080

[18] Mirsal A. Origin, Monitoring & Remediation. In: Soil Pollution. Berlin, Heidelberg: Springer-Verlag; 2008. p. 312. DOI: 10.1007/978-3-540-70777-6

[19] Franco A, Trapp S. Estimation of the soil–water partition coefficient normalized to organic carbon for ionizable organic chemicals. Environmental Toxicology and Chemistry. 2008;**27**:1995-2004. DOI: 10.1897/07-583.1

[20] Sparks DL. 5—Sorption phenomena on soils. In: Environmental Soil Chemistry. 2nd ed. Burlington: Academic Press; 2003. pp. 133-186. DOI: 10.1016/B978-012656446-4/50005-0

[21] Báez ME, Fuentes E, Espinoza J. Characterization of the atrazine sorption process on andisol and ultisol volcanic ash-derived soils: Kinetic parameters and the contribution of humic fractions. Journal of Agricultural and Food Chemistry. 2013;**61**:6150-6160. DOI: 10.1021/jf400950d

[22] Brusseau ML, Famisan GB, Artiola JF, Janick FA, Ian LP, Mark LB. Chemical contaminants. In: Environmental Monitoring and Characterization. Burlington: Academic Press; 2004. pp. 299-312

[23] Caceres-Jensen L, Rodriguez-Becerra J, Escudey M. Impact of physical/chemical properties of volcanic ash-derived soils on mechanisms involved during sorption of ionisable and non-ionisable herbicides. In: Edebali DS, editor. Advanced Sorption Process Applications. London: Intech; 2018. p. 95-149. DOI: 10.5772/intechopen.81155

[24] Tan KL, Hameed BH. Insight into the adsorption kinetics models for the removal of contaminants from aqueous solutions. Journal of the Taiwan Institute of Chemical Engineers. 2017;**74**:25-48. DOI: 10.1016/j.jtice.2017.01.024

[25] Fernández-Bayo JD, Nogales R, Romero E. Evaluation of the sorption process for imidacloprid and diuron in eight agricultural soils from Southern Europe using various kinetic models. Journal of Agricultural and Food Chemistry. 2008;**56**:5266-5272. DOI: 10.1021/jf8004349

[26] Pojananukij N, Wantala K, Neramittagapong S, Lin C, Tanangteerpong D, Neramittagapong A. Improvement of As(III) removal with diatomite overlay nanoscale zero-valent iron (nZVI-D): Adsorption isotherm and adsorption kinetic studies. Water Science and Technology: Water Supply. 2017;**17**:212-220. DOI: 10.2166/ws.2016.120

[27] Valderrama C, Gamisans X, de las Heras X, Farrán A, Cortina JL. Sorption kinetics of polycyclic aromatic hydrocarbons removal using granular activated carbon: Intraparticle diffusion coefficients. Journal of Hazardous Materials. 2008;**157**:386-396. DOI: 10.1016/j.jhazmat.2007.12.119

[28] Worrall F. A study of suspended and colloidal matter in the leachate from lysimeters and its role in pesticide transport. Journal of Environmental Quality. 1999;**28**:595-604. DOI: 10.2134/jeq1999.00472425002800020025x

[29] Thevenot M, Dousset S, Rousseaux S, Andreux F. Influence of organic amendments on diuron leaching through an acidic and a calcareous vineyard soil using undisturbed lysimeters. Environmental Pollution. 2008;**153**:148-156

[30] Zhu YF, Liu XM, Xie Z, Xu JM, Gan J. Metsulfuron-methyl adsorption/desorption in variably charged soils from Southeast China. Fresenius Environmental Bulletin. 2007;**16**:1363-1368

[31] Escudey M, Förster JE, Galindo G. Relevance of organic matter in some chemical and physical characteristics of volcanic ash-derived soils. Communications in Soil Science and Plant Analysis. 2004;**35**:781-797. DOI: 10.1081/css-120030358

[32] Weber JB, McKinnon EJ, Swain LR. Sorption and mobility of 14C-labeled imazaquin and metolachlor in four soils as influenced by soil properties. Journal of Agricultural and Food Chemistry. 2003;**51**:5752-5759. DOI: 10.1021/jf021210t

[33] Piccolo A, Conte P, Scheunert I, Paci M. Atrazine interactions with soil humic substances of different molecular structure. Journal of Environmental Quality. 1998;**27**:1324-1333. DOI: 10.2134/jeq1998.00472425002700060009x

[34] Murano H, Suzuki K, Kayada S, Saito M, Yuge N, Arishiro T, et al. Influence of humic substances and iron and aluminum ions on the sorption of acetamiprid to an arable soil. Science of the Total Environment. 2018;**615**:1478-1484. DOI: 10.1016/j.scitotenv.2017.09.120

[35] Espinoza J, Fuentes E, Báez ME. Sorption behavior of bensulfuron-methyl on andisols and ultisols volcanic ash-derived soils: Contribution of humic fractions and mineral-organic complexes. Environmental Pollution. 2009;**157**:3387-3395. DOI: 10.1016/j.envpol.2009.06.028

Development of Conceptual Model for Eco-Based Strategic Environmental Assessment

Kanokporn Swangjang

Abstract

Since the development of mega projects had been contributed, in consequence, the continuous projects were developed and caused some hidden effects. The main target of this chapter is to develop conceptual model for eco-based strategic environmental assessment (SEA) as the tool to consider the kinetic development resulting from project impacts. Three indicators, namely, environmental assessment, land use, and ecological approach, were selected to support the purpose. For environmental dimension, the contents of Environmental Impact Assessment Guidelines and Environmental Impact Statements were analyzed, using content analysis. Land use change for selected areas was analyzed covering the period of mega project development. For ecosystem, the development of ecological pattern from the past to the present was surveyed and investigated in detail. The results illustrated the hierarchical risk areas from the lowest to the highest. Finally, the conceptual model was developed on the basis of the actual impacts according to the area feature.

Keywords: multiple criteria analysis, ecology, land use, Environmental Impact Assessment, development projects, Thailand

1. Introduction

Since the adoption of the National Environmental Policy Act (NEPA) in the United States in 1969, Environmental Impact Assessment (EIA) has become an increasingly familiar term in many developed and developing countries. International agencies and government worldwide have made considerable progress in requiring the use of EIA for evaluating project proposals [1]. In another view, EIA is a knowledge driven to the following theories in the chain of environmental assessment (EA). Strategic environmental assessment (SEA) is one among them. SEA as one of the series of environmental analysis has played an important role since the middle of the 1970s. The origin of SEA was come from the weak point of EIA as the impact specific for only project level. EIA alone makes insufficient to consider cumulative effect and cannot be used as the direction to clarify the environmental management of overall project [2]. EIA mechanism is the process to assess the consequence and impacts only for project levels, whereas SEA focuses on the consideration of impact on the macro-levels of policy plan and programs. The decision-making of both EIA and SEA is different, depending on the jurisdictions in each country [3]. The development of EIA to the higher level in order to determine and control the impacts from the initial stage of the project decision-making process is essential. Currently, the SEA mechanism is widely used in many countries and international organizations. SEA can operate in various forms and methods, such as SEA for sectorial and regional sections by the World Bank [4]. It is recognized that SEA is one of the key drivers toward the achievement of sustainable development goals.

The extension of the project level (EIA) to the macro-level (SEA) to meet the goals of sustainable development has been conducted by many experts in many regions. The 801 EIA projects in the Czech Republic were evaluated and found the linkage of the project evaluation in EIA follow-up to the SEA [5]. The project level, both positive and negative effects, can be expanded to the policy and planning levels [6]. The setting indicators to study are of primary concerned, depending on the conditions of the study. The classification of indicators influencing a carrying capacity depended on the purpose of application and spatial setting. There are various categories identified by many experts. There are, for example, four components identified, including environmental and ecological, urban facilities, public perception, and institutional categories [7]. Some specific indicators were suggested such as soil, slope, vegetation, wetland, scenic resources, natural hazard, air and water quality, and energy availability; some considered water supply, sewage, waste treatment, railway, road, and housing. These are depended on the purposed of each strategic study.

This chapter aims to illustrate the development of conceptual model of eco-based SEA. The setting of purposes to select the objectives, targets, and indicators was described in Section 2. Section 3 illustrated the case study based on the kinetic development resulting from land use change which brought to consequence ecological impacts. The lesson drawn from the case study leads to the development of conceptual model together with the approach for its fulfillment in Sections 4 and 5, respectively.

2. Eco-based strategic environmental assessment

The setting of objectives, targets, and indicators is necessary for the SEA because the SEA baseline cannot be detailed in-depth, like EIA [8]. Those should be appropriate for the strategic purpose. In order to support the aim to develop eco-based SEA model, the selected factors supporting the purpose are EIA mechanism, land use, and ecological approach. The importance of these can be found from the previous researches, as follows.

2.1 Environmental Impact Assessment mechanism

EIA is an effective tool for managing project life cycles [9]. Research on the mechanism of EIA project began in the early 1980s by studying the role of relevant organizations [10, 11]. The quality of the baseline data that directly concern the selection of environmental components appropriated for such project [12, 13] was important to judge the performance [14]. According to EIA mechanism, EIA follow-up, including monitoring and audit, is the main tool to justify the efficiency of project implementation. Monitoring and audit can be used to measure the actual impact of project activity together with the uncertainty of impact prediction [15]. The study of techniques used to monitor actual impacts during the project operation can suggest some error of impact estimation in EIS, together with the impacts beyond forecasts [16].

The efficiency of project control, including the completion of Environmental Impact Statements (EISs) or EIA reports, the compliance with the conditions of approval, and the factors affecting project decision, was developed during the 1990s [17], together with the suggested criteria to assess the EIA effectiveness [18]. The studies to conduct EIA follow-up was based on the principles of operational phase analysis. The case studies were found in many regions. These examples are the following. The study of factors affected the effectiveness of project monitoring in Australia [19]. A network of components affected the efficiency of the EIA process in Taiwan [20]. The efficiency of the EIA process through the environmental monitoring network, focusing on coastal development projects, was evaluated in Mauritius Island [21]. Similar studies were conducted in Malaysia and Kenya, respectively [12, 17].

The importance of ecological components in the project level as EIA has been realized for a long time. However, it still found problems in terms of perfection and effectiveness, the main reason being due to the methods used for ecological prediction and project management, which was too general, without focusing on the critical issues [22, 23]. However, the relationship among EIA, ecology, and sustainable development is crucial. These were confirmed by many researches [24–26]. All illustrated that EIA can be guided toward sustainable development principles, by extending the scope of social considerations and environment. These combination mechanisms were classified, and some study indicated at least 3 of 14 mechanisms, which are directly related to EIA follow-up during the operational phase [25]. The relationship of social, economic, and ecological variables that contributed to the integration of EIA in sustainable development was also confirmed [27].

The United Nations Environment Programme (UNEP) [28] has established guidelines for monitoring biodiversity given the priority to the ecological level in the ecological monitoring trail. The criteria of UNEP are useful for narrowing ecological index categories and can be used as a guideline for the selection of ecological index at each level in order to track changes in the ecosystem.

2.2 Land use

Change to urban areas has increased significantly in many regions. Land use change is an indicator of ecological change. The loss of green areas resulting from land use change has a further impact on many environmental components. One of those is climate change, the global crisis, which affects biosphere by surface temperature change [29] on both minimum and maximum surface temperatures [30].

Dynamic of land use change is different depending on the kinetic development of each area. The study in Beijing illustrated the severely damaged during 1986–2001 in agricultural areas, due to the indefinite of urban growth [31]. A similar study is found in the suburbs of Bangkok that the pattern of urban land use had been profoundly influenced by past patterns of agricultural land use and landform transformation. The volume of landform transformation occurred over the last half-century had been calculated at $3.2 \times 10^7 \, m^3$, equivalent to 64 km^2 of area flooded to an average depth of 50 cm. This is clear that land use change had occurred in both horizontal and vertical components, which could not be separated from each other [32]. Those lead to the study concerning the arrangement of green areas to limit the future expansion of the city [33]. The approach of land use change could be used to develop an environmental monitoring system [34] and also environmental management by analyzing the pollutant sources from land use classification [35]. Urban Carrying Capacity Assessment System was suggested as an alternative tool for effective urban planning and management [7].

Land use planning based on an ecological network, focusing on biodiversity and the conservation of the habitat from the species level, was recommended [36]. The similar case defined the greenways for land use planning in order to conserve biodiversity in the city area [37]. This is an alternative approach for land use planning to support sustainable purpose. In turn, ecological principles are the basic tool to green areas planned for the city.

2.3 Ecological approach

Relationships between landscape pattern and ecological structure have been widely recognized. Land use change brings to the kinetic development of ecological change. It directly concerns the habitat which is the determining factor for ecosystem component.

The impacts on ecological mechanism are different depending on the purpose. It may be considered in the form of various energy and nutrient cycles and the benefits to humans such as food production or waste treatment system. The ecological mechanism was classified into five categories, including regulation function, habitat function, production function, information function, and carrier function [38].

Any habitat change as one of kinetic development within the ecosystem has an effect on living organisms. Among them, bird is the sensitive organism and detects a change of habitat for us to consider a carrying capacity in the ecosystem. Many researches insisted the impacts of land use change on bird species. The examples are followed. The patterns of habitat change had a significant impact on migratory birds [39]. The study in the twin cities of Minnesota, USA, found different responses of bird community among the rural, the suburb, and the conservative habitats [40]. The research regarding the distance from urban habitat and the road corridor to bird index insisted that urban habitat had not only an effect on the number of birds but also on the species abundance, especially local species [41]. In this research, buffer zone was recommended, at least 400 m from urban area and 300 m from the road. The study at the Island of Damar, the Eastern Indonesia, found the disappearance of bird species due to the expansion of small-scale agriculture. The comparing change of bird group between 1890 and 2001 found the difference of the number for fruit-eating birds and insectivorous birds in different habitat forests [42]. Habitat changes were likely to result in the decline of habitat quality for birds. Such effects occurred especially with birds that consume insects and fruits. This study also provided the characteristics of habitat change. The obvious change from the original forest that affected the new-generation forest was the loss of leaf shade covering, reducing tree height and changing flora types from trees to grass. These factors had significantly resulted in the declining number of fruit-eating birds. The major consequences were the loss and declining number of wild birds. On contrary, the increasing number of birds with opposite behavior, including meadow bird, was common at the same time.

Ecological principles can be applied to manage the landscape as the study in agricultural areas by determining the yield of rice and habitat conservation in the lowlands [43]. Civic engagement was recommended as the essential tool for the resolution of sustainability because eco-civic region can help to understand local people, together with the boundaries of biophysical framework within the actual environment [44].

The relationships of land use and ecology, as reviewed, are closely concerned for both the cause and the effect within each other. The interaction is useful for environmental management based on the carrying capacity of the area. These lead to identify the objectives, the targets, and the indicators to fulfill the development of conceptual model for eco-based SEA.

3. Case study

The case study to support the development of conceptual model for eco-based SEA considered the consequence of mega project and the kinetic development of the surrounding area. Three approaches, including environmental assessment, land use, and ecological principle, were the targets to assess the change within the study area. The selected areas to support the purpose were the areas approached by the airport development, as mega project. These areas are located at the suburb of Bangkok, the capital of Thailand, approximately 700 km^2. According to administrative system, four districts are included, namely, Prawet, Ladkrabang, Bangpli, and Bangsauthong.

Multiple criteria analysis (MCA) is one of decision theories used to justify various factors and conditions to achieve the setting aim. It is suitable for addressing complex aspects with different forms of data in both social and scientific systems. This is done by extending decision to accommodate multi-attributed consequences. This approach is acceptable for SEA in many case studies [45–47]. This case study adapted the main stages of MCA which include the setting goal, the provision of criteria to support the goal, the evaluation of setting criteria, and the direction of ranked alternative. To follow those MCA, the study was divided into four main stages:

(1) To establish the main concept associated with what the study aims.

(2) To set the criteria based on relevant theories. This study deals with three theories, including environmental assessment, land use, and ecological issues.

(3) To identify the indicators for each established criteria. These are the variables used in the decision making.

(4) To determine the direction of the variable, by ranking the status of each variable setting from the highest to the lowest.

Stage 1 is the setting of the main purpose. The selected criteria, in stage 2, are based on the circumstance of the areas and their kinetic development; as to the case study, review literature of previous researches was supported. The selected criteria were in-depth investigated and detailed in Section 3.1. These were the baseline to assess the SEA for stages 3 and 4 in Section 3.2. The development of conceptual model of eco-based SEA was clarified in Section 4.

3.1 Results of the case study

Study methods for each set of the criteria, including environmental assessment, land use, and ecological approaches, were appropriately conducted to support the framework of eco-based SEA. The results were shown in **Table 1**. Again, it should be noted that this model is one of the cases from a tropical country under the conditions of mega project development.

3.2 Integration to strategic environmental assessment

The imbalance between the development and the conservation was found from the results of the case study. Some effective tool toward sustainable achievement was required. Among those, SEA is one. The integration of the case study with SEA was conducted by programmatic SEA model since the groups of projects were analyzed in the same boundary area [46]. Hence, the specification of "SEA requirement of project activities" was the first screening process in order to select only the significant activities included in the SEA. The legislation, the Town and Urban Planning, the characteristics of the area, and others were the factors to support this eco-based SEA.

Strategic ecological assessment included the following stages: the scope for analysis, the prediction of future change, the alternative consideration, and the control approach. These can be described as follows:

Study approaches	Results
Environmental assessment; the roles of competent agencies	
Analysis of law and regulation of competent agencies regarding the contents of project control	The environmental control mechanism of different agencies found some question regarding their purposes and collaboration
Environmental assessment; guidelines quality	
Content analysis of EIA guidelines focusing on ecological issues, including: (1) General guideline (2) Project-specific guidelines including: - Airport project - Housing project - Transportation project - Power plant project - Petroleum and oil pipeline project - Industrial project Review criteria were developed. The content in guidelines according to the setting criteria was scored through their quality	*Baseline study*: the specification of boundary of study, focusing on impact area, and method of ecological study was sufficient for the guidance of EIA study. However, general details were found for data analysis and presentation *Impact assessment*: the guidance for impact coverage project life cycle was sufficient; however, the depth details for ecological impact analysis were inadequate *Mitigation and monitoring measures*: the guidelines supported standard format for program presentation. Ecological aspect for program identification was presented only through airport project guideline The score values of EIA guideline content, according to the parts of EIA study, from the highest to the lowest quality were monitoring, mitigation, impact assessment, and baseline study, accordingly
Environmental assessment; EIS quality	
Content analysis of ecological detailed in EISs, including: (1) Airport project and related projects (2) Infrastructure projects (3) Other projects within study area The sets of review criteria, which are different from the guidelines were developed. The quality of EIS response to each review criterion was scored	Ecological details were mostly presented in the stage of baseline study, followed by impact assessment Negligible details were found in mitigation and monitoring. As to their quality, the linkage of ecological factors in baseline detail was weak. In the following stages, impact assessment, the results of ecological baseline were scarcely considered to assess the impacts These bring to the unclear impact direction, especially ecological mechanism within the study area. Ecological mitigation and monitoring identification were not concurred with the result of impact assessment
Environmental assessment; monitoring efficiency	
Two groups of development projects, including: (1) Project that required EIA (2) Project that did not require EIA (industrial projects) These are conducted by: (1) The content analysis of monitoring EISs (only for EIA projects) (2) The investigation of monitoring compliance by auditing the monitoring reports (3) The consistence between project location and the Town and Urban Planning by overlay mapping	In comparison, projects that required EIA were predominant, as follows: - Monitoring details fulfilled the aspects of environmental components, monitoring frequency, and stations which were specific for each project feature However, these lead to the increase in monitoring cost compared with project that did not require EIA - The enforcement by competent agencies was strengthening in terms of the linkage of monitoring performance - Project setting complied with the Town and Urban Planning - The weakness monitoring, especially ecological aspects, was found for projects that required EIA, and the missing was found for projects that did not require EIA
Land use; overall study areas	

Study approaches	Results
Land use change was done by GIS layer interpretation of the study area during the year before (1994) and after (2002) airport project development Land use was grouped into three types, including: (1) Development area (2) Semi-developed area (3) Conservative area	- The increase in the development area was prominent, with 40%. Semi-developed area was more or less, with 37%. Insignificant change was found for the conservative area, with 0.05 - The kinetics of land use change was caused mainly by transportation network, which leads to the increase in housing projects and recruited the increase of population. These areas were especially the area around airport development
Land use; the pattern of significant project change (housing project)	
The expansion of housing projects was conducted by satellite interpretation in the years 1981, 1987, 1996, 2002, and 2006 and overlaid with the map of the Town and Urban Planning	The expansion of housing projects had been rapidly increased in the years 1981, 1987, 1996, 2002, and 2006. The increase (%) was: Prawet 4.39, 8.27, 9.66, 13.66, and 22.15 Ladkrabang 0.31, 1.59, 4.03, 5.18, and 8.97 Bangpli 0.42, 1.32, 3.69, 5.08, and 8.95 Bangsauthong 0, 0.75, 1.01, 1.29, and 1.91 According to the results, housing project after the year 2002, in which the airport initially operated, was sharply increased. Only 13% of these housing projects were required EIA, according to the Thai's EIA legislation. Significantly, 4.5% of EIA housing projects conducted monitoring performance Regarding the condition of the Town and Urban Planning, it was found that: - Location of housing projects was mostly in medium-density dwelling stipulated area - The expansion of the housing projects encroached 13.40% (in Ladkrabang) of conservative urban area, whereas the provisions of the Ministerial Regulations of the Town and Urban Planning Act enforce not 10% exceeding - 7.91% of housing projects in Bangpli were located in the industrial area. These reflect to the risk impacts of the projects themselves - The expansion of housing projects was inconsistent with the rate of population increase
Ecological approach; the change of local species	
The study was conducted through: (1) Questionnaire interview to local people (2) Bird count surveys in the designed land use	The pattern of land use change was the main factor. Originally, paddy fields were dominant in the area. After the airport development, the pattern of land use change can be divided into two groups, as follows: (1) Paddy fields to fish farms and to urban area (2) Paddy fields to wilderness and to urban development These affect the change of local species including: - Species disappearance, both in the stages of paddy fields to fish farms and fish farms to urban area - The increasing number of urban species, especially for bird species - The change of species behavior, such as from migratory birds to permanent local birds - The change of ecological index, including species diversity, abundance index, and similarity index The highest values of ecological diversity were found in paddy fields. Local species were significantly changed, especially in the stage from paddy fields to fish farms

Sources: [48–50]

Table 1.
The stages and results of the case study.

Step 1: Determining the scope of strategic environmental analysis

Since there are many conditions to analyze eco-based SEA, the identification of aspects, targets, objectives, and indicators is important. The results of the case study were integrated with the SEA theory [46] to determine the relevant variables. Targets define issues that are likely the impact; objectives are the desired change that should be consistent with the target. Indicators are the variables that represent the direction of change (**Table 2**). These factors are important in considering basic environmental information to support conceptual approach.

Purposes	Targets	Objectives	Indicators
Land use	- Project expansion complied with the provision of the Town and Urban Planning Act - The consideration of ecological aspects in any development projects	- The expansion of the project in the agricultural conservation area and rural agriculture area, not exceeding the requirements in the Town and Urban Planning Act (the stipulation is less than 10%) - The growth of significant development projects is controlled	The audit of incremental rate for development projects meets the requirement of the Town and Country Planning
Environmental Impact Assessment	- Project that required EIA - Project that did not require EIA - Ecological issues in Environmental Impact Assessment	*Project that required EIA* The requirements are: - The guidance of ecological issues in EIA guidelines - The appropriateness of ecological impact study - The relationships of ecological contents in guidelines and EISs - The importance of environmental control mechanism, especially mitigation and monitoring during project implementation	- Ecological issues in EISs are qualified with the criteria established - The number of development projects to meet the requirement of the conditions of approval (mitigation and monitoring measures) is examined
		Project that did not require EIA - Environmental monitoring is the priority as a tool to control projects - Project control mechanism embraces more collaborative and inclusive environmental concern by relevant agencies	- The number of development projects to meet the requirements of monitoring implementation is examined - Monitoring details investigate the effectiveness
Ecological approach	Maintaining biodiversity and local species within the area	The habitats for local species are preserved, with appropriate types and size	- The appropriateness of ecological index and species types are identified and monitored to warn the ecological change

Table 2.
The identification of targets, objectives, and indicators.

Step 2: Future change without control mechanism

Baseline data for the strategic level should not provide definite details, like the project EIA level [46]. From **Table 2**, the baseline data were established, following three main areas, including the change of land use, projects enforced by EIA, and local ecosystem. The identification of the conditions in such areas and the future trends in case of lack of any control mechanism were presented in **Table 3**. The limit of the integrity of the environmental database is the obstacles in some countries, like this case study. Therefore, the appropriate analysis corresponding to the area is necessary for the future trends of a specific area. The environmental trends are variable factors used as the baseline to determine any change of the indicators considered [46].

Step 3: Alternative consideration

Alternative identification is crucial for SEA. The example provided in **Table 4** was the result from the case study. Alternative conditions in each area were differed,

Purposes	Limitations	Future trends without control mechanism
Land use	- The development of mega projects taking into account economic outcome was the first priority - The expansion of housing projects allocated in the areas that conflict with the Town and Urban Planning, especially in the conservation -agriculture area and rural-agriculture area	- Project expansion will lack control mechanism, especially for the projects, which are unclearly enforced by competent agencies - The expansion of housing projects will over the requirement of the Town and Urban Planning in the conservative area - The increase of urban area will be opposite to the green area
Environmental Impact Assessment	- The proportion of the number of projects that required EIA was minimal compared with all developmental projects Hence, EIA was not the main tool to control impacts of project activities - The quality of ecological contents in EIA guidelines and EISs is still in question - Ecological mitigation and monitoring as the conditions of approval from EIA studies were missing, which further affected ecological control during project implementation	- Ecological issues will be overlooked unless the mechanism to stimulate is sufficient - The importance of EIA declines, whereas SEA cannot be replaced unless defined in the highest legal hierarchy
Ecological approach	- Land use was an important factor for the development of infrastructure within the ecosystem	- Habitat change will be the main factor that affects species types and ecosystems as a whole. The change is due to the urbanization together with the decline of green space. These are the results of the increase of development projects and the decline of green area itself - The change of local ecology will affect ecosystem in macro-level - The target of sustainable development could not be achieved due to the focus only on economic factors, without ecological values

Table 3.
Environmental baseline for the strategic level.

Study areas	Kinetic conditions	Factors to be considered as appropriate alternatives
Prawet	- Urbanization rate was high - The number of housing projects has dramatically been increased. Among these, only few are required EIA - According to the Town and Urban Planning, residential areas are defined as more than 30%. This condition was the limiting factor for the ecological considerations	- EIA mechanism requires more rigorous tool for projects that required EIA - As to projects that did not require EIA, the alternative controls should be enforced by the competent agencies
Ladkrabang	- The housing projects have been expanded in green belt area - The expansion of projects that did not require EIA is limitless	The rural and agricultural conservation areas, which are the city's prosperity to the green area, are the priority to allow any the development projects
Bangpli	- The expansion of housing projects encroached the industrial setting area and went over the limit of rural and agricultural areas - The change of local ecosystem was caused by land use diversity within the area	- The screening of development projects in accordance with the Town and Urban Planning should be the first concern
Bangsauthong	- The change of agriculture types was dominant - Development projects were controlled in the low level	- The alternative methods of project control should be the first concern in this area

Table4.
Kinetic conditions considered as alternatives.

based on the multiple criteria analysis, which provided the score ranking for each factors. The alternative appropriateness in each area should take into account the nature of the development projects within the areas together with the kinetic conditions of the development. Alternative consideration based on existing constraints directly concerns the scope of activity frameworks under sustainable development.

Step 4: Impact assessment

Impact assessment includes impact prediction and evaluation. The methods used are varied depending on the appropriateness. The baseline in **Table 3** and the conditions of alternative in **Table 4** were assessed the impacts. The results of this stage provide the overall possibility of change. This stage is different from the assessment of EIA level which is the proactive assessment. As to the SEA level, the assessment is conducted after the operation of activities in order to find out their future trends.

Step 5: Monitoring

Monitoring of indicators specified is important for SEA in order to detect any environmental change resulting from activities considered for each area. The factors to identify should include:

- The policy support

- The coverage of environmental constraints within the area

- The appropriateness of parameters selection in terms of the budget and its benefit

- The capability to detect any change within the area

- The efficiency to identify and decide the priority of environmental conditions

- The resilience for any unexpected conditions

4. Conceptual model of eco-based strategic environmental assessment

The aims of the SEA [46, 51, 52], focusing on specific ecological issues resulting from the case study, lead to the proposed conceptual model of the eco-based SEA in **Figure 1**.

The relationship of the main factors affecting the environment in the area is presented. At policy and planning levels, legal framework (No. 1) sets the direction of activities at the program and project levels. The Town and Urban Planning (No. 2) is a key factor to scope any development activities in each area. The change of land use is caused by two parts. The first part is due to development projects (No. 4), projects that required EIA (No. 5) and projects that did not require EIA (No. 6). These projects require official monitoring mechanisms and the audit from the competent agencies. The second part is due to the other local activities (No. 7) such as the change in agricultural types within the green area. Land use change caused by project activities can be controlled by the Town and Urban Planning, while another is caused by economic outcome and the unseen disaster.

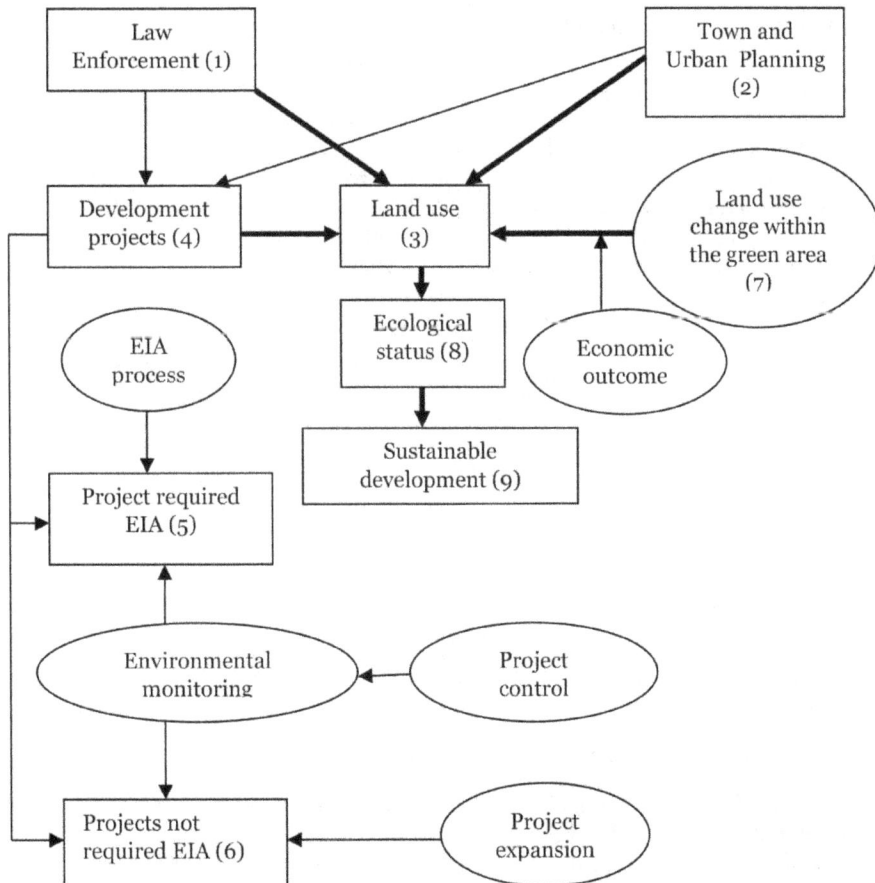

Figure 1.
Conceptual model of eco-based strategic environmental assessment.

Three aspects are raised from the conceptual model, including projects that required EIA and projects that did not require EIA, and the change of land use within green areas. These are discussed as follows.

Projects that required EIA: the main factors are the environmental impact study and environmental quality monitoring.

Ecological issues in EIA guidelines have a direct effect on the details in the EISs. The quality of data in one step will affect another. It seems that EIA is a satisfactory tool for identifying the adverse impact of projects and, consequently, monitoring the administrative procedures of government agencies. The environmental studies reported in an EIA are detailed and specific to the individual project. Furthermore, the prescriptions to reduce the impact that raised from a project are the mitigation and monitoring programs included in an EIS as project control mechanism.

The achievement of mitigation and monitoring depends on several factors including (1) the compliance by the project proponents. This is due to the details contained in the measures that encourage the performance and (2) the control by relevant agencies. This is depending on the legislation of the respective agency. It is essential that the regulations of the relevant agencies require the concurrence with the EIA legislation. A definition and allocation of roles and responsibilities to cover the requirements of follow-up activities among all key actors are required.

Projects that did not require EIA: the main factors are the Town and Urban Planning, project controlled by competent agencies and project expansion.

Monitoring performance of projects that did not require EIA depends on the requirements of competent agencies. The normal practice is that, for one type of project, only a particular suit of issue will be considered. In effect, these issues reflect the legal responsibilities of the agency based on past experience.

Another question concerns the expansion of projects that defined as non-severe impacts, especially housing project. The finding from the case study was that only 12% of the total required EIA and among 4.5% of these conducted monitoring performance. It seems that environmental control mechanism of these projects was too weak. The Town and Urban Planning is another tool to control; however, the unlimited expansion of housing projects was found in some restricted areas. These are crucial factors contributing to ecological change.

The change of land use within green areas: the change within green areas due to economic outcome is another hiding factor affecting ecological change. The factors causing these changes are difficult to control. It is a silent disaster that causes kinetic ecological change. The example of case study clearly showed that the change from paddy fields to fish farms affected species, habitat, and ecological mechanisms (No. 8), one of the sustainable approaches (No. 9).

To sum up, the relationships of eco-based SEA are depended on three components, including:

(1) Land use: the main factor is No. 3, with relevant elements (Nos. 1, 2, 4, and 7).

(2) Environmental assessment: the main factor is No. 4, with relevant elements (Nos. 5 and 6).

(3) Ecosystem: the main factor is No. 8, with inputs (Nos. 1–7) and output (No. 9).

The main conceptual model has been expanded to sub-frameworks, focusing on development projects, in **Figure 2**. The main factors of this sub-model are the Town and Country Planning due to its enforcement to specify land use development within the area and the legal enforcement by competent authorities.

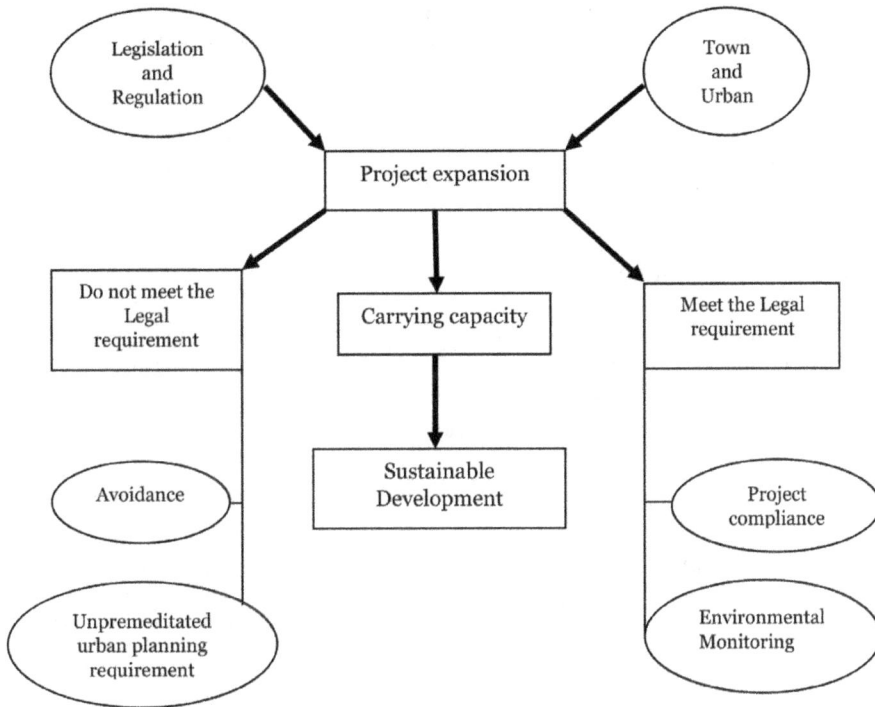

Figure 2.
Sub-conceptual model of development project expansion.

The expansion of two main types of development projects, mentioned previously, directly affects land use change within the area. The characteristics of change can be divided into two groups, according to the compliance with the legal specification.

The case does not meet the requirements of the law: the causes are followed. The first cause is avoidance, with emphasis on real estate development projects. Indeed, both housing projects and industrial projects are in this condition. However, the enforcement by control agencies is somewhat different. Industrial projects are controlled by Department of Industrial Works which has a strict control mechanism, whereas the unclear environmental control agency is put into housing projects. The second cause is unpremeditated which is mainly caused by the expansion of the project beyond the land use requirement specified in the Town and Urban Planning.

The case is in accordance with the requirements of the law: there are two factors concerned. The first factor is the project location that meets the requirements of the Town and Urban Planning. However, negative effects cannot guarantee for this group without the effective monitoring mechanism. The results of the case study were found that the land use regulations affect the slowdown of new real estate development. It seems that urban expansion is somewhat beneficial for green area preservation. The second factor is the environmental impact monitoring of the projects. This is an important mechanism to control the environmental impact from project activities. The lesson was learned from the case study that only 22.67 and 20.52% of projects that required and did not require EIA, respectively, were performed. Notably, in compliance group, the performance was inefficient.

Land use change directly effects on the appearance of ecological status. It is the crucial factor for the achievement of project activity control. Is it sustainable? For example, the agricultural changes directly affect kinetic change in species, confirmed by the case study result. This is beyond the control of the Town and Urban Planning since the activity continues to be classified as green! But issues need to be realized how these areas are not being compromised by the legal enforcement from the activities of some development projects. The study was found that green belt area was affected by urban expansion, with more than 10% as defined in the Town and Urban Planning. Therefore, the expansion of development projects should be concerned and rigorous by the relevant agencies.

5. The approach for ecological fulfillment

Integrating ecological issues into the environmental impact study (**Table 5**) was crucial to achieve the model setting. It could be channeled into projects that are

Steps	The integration of ecological issues	The enhancement of relationship between EA guidelines and EISs
Ecological level	- Ecological impact should be cleared at all stages of the environmental impact study. Such eco-level considered is appropriate with project activities and the features of their location - Biodiversity should be focused by consideration on ecosystem integrity from the relationship between project and site development, which is a determinant of habitat and ecosystems - The flow of ecological details should be balanced at all steps of impact study	- The integration of ecological details in each level of environmental impact study should be taken into account the budget and time constraints - The establishment of review criteria is required in EIS submission process
Ecological baseline	- The scope of ecological study should be comprehensive and flexible, based on the feasibility of the impacts - Ecological information should cover the space and period of impact possibility - The formal guidance should provide the clarification of the biodiversity and the minimum requirement for the direction of EIA study - A comprehensive study of each ecological level for such issue should be based on project details and location - The linkage of ecological baseline and its impact assessment is required	- The role of expert committee in EIS submission is significant for the quality of ecological impact assessment - The linkage of ecology and EIA disciplines is required for appropriate integration - Ecological information to be used as a basis for the assessment of ecological impacts should be emphasized
Ecological assessment	- Ecological impact assessment should be based on ecological baseline - The guidelines should set the criteria for ecological impact analysis, focusing on biodiversity issues - Integration of biodiversity issues should consider the coverage of ecological details and their flexibility depending on project conditions and location - The flexibility of techniques and methods used to identify and analyze impacts under the principle of sustainable development is required	- The agreement of responsible agencies in EIA process is essential for project performance - The role of all agencies concerned during EIA process and EIA follow-up is necessary for the quality of EIS and the efficiency of project implementation
Ecological mitigation and monitoring	- The set of criteria for mitigation/monitoring measures is required, with ecological standard based on clear references - Mitigation/monitoring identification should be done by prioritizing the significance of ecological impacts	- EIA guidelines should provide the definite mitigation/monitoring identification - The essential role of project control agencies and project proponents is necessary to support and control project performance

Table 5.
The integration of ecological issues in EIA study.

subject to EIA through the mechanism of SEA. The importance of EIA guidelines is a fundamental tool in studying the ecological impact. The accuracy and appropriateness of baseline data are prominent. Data presentation should be appropriate, not too short or too long to identify their subsequent impacts. Biodiversity is firstly considered for ecological information in order to understand the overall ecological pattern within the area. The composition of the ecological level should be of great importance, such as the indicators species, the relationships between local and regional factors, the species of habitats [40, 53], the habitat loss, and the change of species distribution [54–56]. Quantitative approach is the possibility to integrate ecological science in the environmental impact study by the consideration of the variation of species, extent, and timing [39, 54].

In the process of EIA study, the impact of project activities to any kinetic habitat change should be highlighted because it is the main cause to the change of ecosystem composition, especially the change to species index [56, 57]. Ecological impact study should be conducted based on cognitive theoretical knowledge [58]. Drawing these theories together with the details of the project is very important and that is often overlooked. Good ecological baselines together with the minimal error of ecological impact study directly satisfy mitigation and monitoring measures. The reflect mitigation and monitoring can be examined through the possibilities of biodiversity change due to project activities. The concern agencies are crucial to enforce the project implementation as a result of environmental impact studies.

6. Conclusion

Eco-based SEA model here was developed from the case study derived from mega project development, which both direct and indirect effects on complex conditions, finally, to ecosystem which is one of the key indicators in sustainable development. When each issue was pinpointed, the main cause of impacts within the area was not only from the established mega project but also from the change of continuous activities. The kinetic changes due to development projects, themselves, and the kinetic changes due to land use pattern in the same group, particularly the change within agricultural areas from paddy fields to fish farms, were included.

From the three dimensions of model, these were EIA, land use, and ecology to support the setting purpose focusing on ecological issues. The integration of existing strategies and the results of the case study could be adapted for the appropriateness of the area. Ecological outcomes were considered as a result of activities within such area and the status of the area to support any activities. The conceptual model clearly illustrates in three cognitive, in particular their relationships. All three variables were integrated into SEA in accordance with the limitations of each area, focusing on the priority of ecosystem.

In summary, the model illustrates the importance of considering environmental issues as a whole from their cause to the final output. That is the kinetic ecological change. It can answer the question of large-scale project development, which is a continuation of the macro-level. Is in line of the sustainable development approach?

Acknowledgements

This article is extracted from the project "The development of mechanism for strategic ecological environmental assessment; Suvarnabhumi Airport case study" which is funded by the Office of the Commission for Higher Education and the Thailand Research Fund. It is also partially supported by the Faculty of Science, Silpakorn University.

Conflict of interest

The author would like to declare that there are no conflicts of interest for the entirety of this text.

Author details

Kanokporn Swangjang
Department of Environmental Science, Faculty of Science, Silpakorn University, Thailand

*Address all correspondence to: swangjang_k@su.ac.th

IntechOpen

References

[1] Wathern P. An introduction guide to environmental impact assessment. In: Wathern P, editor. Environmental Impact Assessment: Theory and Practice. London: Unwin Hyman; 1988. pp. 3-30

[2] Dixon J, Therivel R. Managing cumulative impacts: Making it happen. In: Sadler B, Aschemann R, Dusík J, Fischer TB, Partidario M, Verheem R, editors. Handbook of Strategic Environmental Assessment. Oxford: Earthscan; 2011. pp. 380-394

[3] Sadler B. Taking stock of SEA. In: Sadler B, Aschemann R, Dusík J, Fischer TB, Partidario M, Verheem R, editors. Handbook of Strategic Environmental Assessment. Oxford: Earthscan; 2011. pp. 1-20

[4] World Bank. Strategic Environmental Assessment. 2013. Available from: www.worldbank.org/en/topic/environment/brief/strategic-environmental-assessment [Accessed: 2018-09-11]

[5] Branis M, Christopoulos S. Mandated monitoring of post-project impacts in the Czech EIA. Environmental Impact Assessment Review. 2005;25:227-238. DOI: 10.1016/j.eiar.2004.09.001

[6] Connelly S, Richardson T. Value-driven SEA: Time for an environmental justice perspective. Environmental Impact Assessment Review. 2005;25:391-409. DOI: 10.1016/j.eiar.2004.09.002

[7] Oh K, Jeong Y, Lee D, Choi J. Determining development density using the urban carrying capacity assessment system. Landscape and Urban Planning. 2005;73:1-15. DOI: 10.1016/j.landurbplan.2004.06.002

[8] Donnelly A, O'Mahony T. Development and application of environmental indicator in SEA. In:

Sadler B, Aschemann R, Dusík J, Fischer TB, Partidario M, Verheem R, editors. Handbook of Strategic Environmental Assessment. Oxford: Earthscan; 2011. pp. 338-355

[9] Canter LW. Environmental Impact Assessment. Singapore: McGraw Hill; 1996. 374 p

[10] Exner KK, Nelson NK. Environmental follow-up to assessment and mitigation for construction in Alberta. In: Sadler B, editor. Proceeding of the Conference on Follow-up/Audit of EIA Results; October 1985; Banff Centre. pp. 470-483

[11] McCallum DR. Environmental follow-up to federal projects: A national review. In: Sadler B, editor. Proceeding of the Conference on Follow-up/Audit of EIA Results; October 1985; Banff Centre. pp. 163-173

[12] Said AM. The Practice of Post-Monitoring and Audit in Environmental Impact Assessment in Malaysia [Thesis]. United Kingdom: University of Wales, Aberystwyth; 1997

[13] Culhane PJ. Post-EIS environmental auditing: A first step to making rational environmental assessment a reality. The Environmental Professional. 1993;15:66-75

[14] Brew D, Lee N. Reviewing the quality donor agency environmental assessment guidelines. Project Appraisal. 1996;11:79-84. DOI: 10.1080/02688867.1996.9727022

[15] Domeney R. Project management and team operation in environmental impact assessment [thesis]. United Kingdom: University of Wales, Aberystwyth; 1996

[16] Wood C. Assessing techniques of assessment: Post-development

auditing of noise predictive schemas in environmental impact assessment. Impact Assessment and Project Appraisal. 1999;**17**(3):217-226. DOI: 10.3152/147154699781767828

[17] Hirji R, Ortolano L. EIA effectiveness and mechanisms of control: Case studies of water resources development in Kenya. International Journal of Water Resources Development. 1991;**7**(3):154-167. DOI: 10.1080/07900629108722508

[18] Wood C, Bailey J. Predominance and independence in environmental impact assessment: The western Australian model. Environmental Impact Assessment Review. 1994;**14**:37-59. DOI: 10.1016/0195-9255(94)90041-8

[19] Buckley R. Auditing the precision and accuracy of environmental impact assessment in Australia. Environmental Monitoring and Assessment. 1991;**18**:1-23

[20] Leu WS, Williams WP, Bark WA. Development of an environmental impact assessment evaluation method and its application: Taiwan case study. Environmental Impact Assessment Review. 1996;**16**:115-133. DOI: 10.1016/0195-9255(95)00107-7

[21] Ramjeawon T, Beedassy R. Evaluation of the EIA system on the Island of Mauritius and development of an environmental monitoring plan framework. Environmental Impact Assessment Review. 2004;**24**:537-549. DOI: 10.1016/j.eiar.2004.01.001

[22] Wathern P. Ecological impact assessment. In: Petts J, editor. Handbook of Environmental Impact Assessment. Oxford: Blackwell; 1999. pp. 327-346

[23] Joao E. How scale affects environmental impact assessment. Environmental Impact Assessment Review. 2002;**22**:289-310. DOI: 10.1016/s0195-9255(02)00016-1

[24] Devuyst D, Hens L. Introducing and measuring sustainable development initiatives by local authorities in Canada and Flenders (Belgium). Environment, Development and Sustainability. 2000;**2**:81-105

[25] Scrase I, Sheate W. Integration and integrated approaches to assessment: What do they mean for the environment? Journal of Environmental Policy and Planning. 2002;**4**(4):275-294. DOI: 10.1002/jepp.117

[26] Marsden S, Dovers S. Strategic Environmental Assessment in Australasia. Sydney: The Federation Press; 2002. 219 p

[27] Pope J, Annandale D, Morrison-Sauders A. Conceptualising sustainability assessment. Environmental Impact Assessment Review. 2004;**24**:595-616. DOI: 10.1016/j.eiar.2004.03.001

[28] United Nations Environmental Programme (UNEP). Convention on Biological Diversity. Nairobi: UNEP; 1992

[29] Wagner M. Assessment of the environmental consequences of infill development [thesis]. Germany: Munich Technical University; 1992

[30] Pauliet S, Ennos R, Golding Y. Modeling the environmental impacts of urban land use and land cover change—A study in Merseyside, UK. Landscape and Urban Planning. 2005;**71**:295-310. DOI: 10.1016/j.landurbplan.2004.03.009

[31] Wu Q, Li HQ, Wang RS, Paulussen J, He J, Wang M, et al. Monitoring and prediction land use change in Beijing using remote sensing and GIS. Landscape and Urban Planning. 2006;**78**:322-333. DOI: 10.1016/j.landurbplan.2005.10.002

[32] Hara Y, Takeuchi K, Okubo S. Urbanization linked with past

agricultural landuse patterns in the urban fringe of a Deltaic Asian Mega-City: A case study in Bangkok. Landscape and Urban Planning. 2005;**73**:16-28. DOI: 10.1016/j. landurbplan.2004.07.002

[33] Li F, Wang R, Paulussen J, Liu X. Comprehensive concept planning of urban greening based ecological principles: A case study in Beijing, China. Landscape and Urban Planning. 2005;**72**:325-336. DOI: 10.1016/j. landurbplan.2004.04.002

[34] Olsen LM, Dale VH, Foster T. Landscape patterns as indicators of ecological change at Fort Benning, Georgia, USA. Landscape and Urban Planning. 2007;**79**:137-149. DOI: 10.1016/j.landurbplan.2006.02.007

[35] Park M, Stenstrom MK. Classifying environmentally significant urban land uses with satellite imagery. Journal of Environmental Management. 2008;**86**:181-192. DOI: 10.1016/j. jenvman.2006.12.010

[36] Opdam P, Steingrover E, Rooij SV. Ecological networks: A spatial concept for multi-actor planning of sustainable landscapes. Landscape and Urban Planning. 2006;**75**:322-332. DOI: 10.1016/j.landurbplan.2005.02.015

[37] Bryant MM. Urban landscape conservation and the role of ecological greenways at local and metropolitan scales. Landscape and Urban Planning. 2006;**76**:23-44. DOI: 10.1016/j. landurbplan.2004.09.029

[38] Guillermo AM, Macoun P. Guidelines for Applying Multi-Criteria Analysis to the Assessment of Criteria and Indicators. Jakarta: Centre for International Forest Research; 1999. 82 p

[39] Sirami C, Lluis B, Burfield I, Fonderflick J, Martin JL. Is land abandonment having an impact on biodiversity? A meta-analytical approach to bird distribution changes in the North-Western Mediterranean. Biological Conservation. 2008;**141**:450-459. DOI: 10.1016/j.biocon.2007.10.015

[40] Chapman AK, Reich BP. Land use and habitat gradients determine bird community diversity and abundance in suburban, rural and reserve landscape of Minnesota, USA. Biological Conservation. 2006;**135**:527-541. DOI: 10.1016/j.biocon.2006.10.050

[41] Palomino D, Carrascal LM. Threshold distance to nearby cities and roads influence the bird community of a mosaic landscape. Biological Conservation. 2007;**40**:100-109. DOI: 10.1016/j.biocon.2007.07.029

[42] Trainor RC. Change in bird species composition on a remote and well-forested Wallacean Island, South-East Asia. Biological Conservation. 2007;**140**:373-385. DOI: 10.1016/j. biocon.2007.08.022

[43] Musacchio LR, Coulson RN. Landscape ecological planning process for wetland, waterfowl, and farmland conservation. Landscape and Urban Planning. 2001;**56**:125-147. DOI: 10.1016/s0169-2046(01)00175-x

[44] Brunckhorst D, Coop P, Reeve I. Eco-civic optimisation: A nested framework for planning and managing landscape. Landscape and Urban Planning. 2006;**75**:265-281. DOI: 10.1016/J.landurplan.2005.02.013

[45] Groot R. Function analysis and valuation as a tool to access land use conflicts in planning for sustainable, multifunctional landscape. Landscape and Urban Planning. 2006;**75**:175-186. DOI: 10.1016/j.landurbplan.2005.02.016

[46] Therivel R. Strategic Environmental Assessment in Action. London: Earthscan; 2004. 272 p

[47] Schmidt M, Storch H, Helbron H. SEA for agricultural programmes in

the EU. In: Schmidt M, Joao E, Albrecht E, editors. Implementation Strategic Environmental Assessment. Berlin: Springer; 2005. pp. 599-620

[48] Swangjang K, Iamaram V. Change of land use patterns in the area close to the airport development area and some implicating factors. Sustainability. 2011;**3**:1517-1530. DOI: 10.3390/su3091517

[49] Swangjang K. Ecological impact behind mega project development. International Journal of Environmental Science and Development. 2015;**6**(8):620-624. DOI: 10.7763/IJESD.2015.V6.669

[50] Swangjang K. Ecological Impact Assessment; Relationships of Environmental Impact Studies. Germany: Lambert; 2017. 71 p

[51] Sadler B, Verheem R. Country Status Reports on Environmental Impact Assessment: Results of an International Survey. Utrecht: EIA Commission; 1996

[52] Partidario MR. Strategic environmental assessment: Principles and potential. In: Petts J, editor. Handbook of Environmental Impact Assessment. Oxford: Blackwell; 1999. pp. 380-409

[53] Devictor V, Jiguet F. Community richness and stability in agricultural landscapes: The importance of surrounding habitats. Agriculture, Ecosystem & Environment. 2007;**120**:179-184. DOI: 10.1016/j.agee.2006.08.013

[54] Gontier M. Scale issue in the assessment of ecological impacts using a GIS-based habitat model—A case study for the Stockholm Region. Environmental Impact Assessment Review. 2007;**27**:440-459. DOI: 10.1016/j.eiar.2007.02.003

[55] Fuller RM, Devereux BJ, Gillings S, Hill A, Amable GS. Bird distributions relative to remotely sensed habitats in Great Britain: Towards a framework for national modeling. Journal of Environmental Management. 2007;**84**:586-605. DOI: 10.1016/j.jenvman.2006.07.001

[56] Mortberg UM, Balfors Knol WC. Landscape ecological assessment: A tool for integrating biodiversity issues in strategic environmental assessment. Journal of Environmental Management. 2007;**82**:457-470. DOI: 10.1016/j.jenvman.2006.01.005

[57] Thompson GG. Terrestrial vertebrate fauna surveys for the preparation of environmental impact assessments; how can we do it better? A Western Australian example. Environmental Impact Assessment Review. 2007;**27**:41-61. DOI: 10.1016/j.eiar.2006.08.001

[58] Hiddink JG, Jennings S, Kaiser MJ. Assessing and predicting the relative ecological impacts of disturbance on habitats with different sensitivities. Journal of Applied Ecology. 2007;**44**:405-413. DOI: 10.1111/j.1365-2664.2007.01274.x